ENERGY AND INFRASTRUCTURE

Volume 7

Nuclear Juggernaut
The transport of radioactive materials

Full list of titles in the set
ENERGY AND INFRASTRUCTURE

Nuclear Juggernaut
The transport of radioactive materials

Martin Bond

from Routledge

First published by Earthscan in the UK and USA in 1992
First edition 2009

For a full list of publications please contact:
Earthscan
2 Park Square, Milton Park, Abingdon, Oxfordshire OX14 4RN
Simultaneously published in the USA and Canada by Earthscan
711 Third Avenue, New York, NY 10017

First issued in paperback 2016

Earthscan is an imprint of the Taylor & Francis Group, an informa business

A catalogue record for this book is available from the British Library

Library of Congress Cataloging-in-Publication Data has been applied for

Publisher's note
The publisher has made every effort to ensure the quality of this reprint, but
points out that some imperfections in the original copies may be apparent.

At Earthscan we strive to minimize our environmental impacts and carbon
footprint through reducing waste, recycling and offsetting our CO_2
emissions, including those created through publication of this book. For
more details of our environmental policy, see www.earthscan.co.uk.

ISBN 13: 978-1-138-99454-6 (pbk)
ISBN 13: 978-1-84407-979-7 (hbk)
ISBN 978-1-84407-972-8 (Energy and Infrastructure set)
ISBN 978-1-84407-930-8 (Earthscan Library Collection)

Martin Bond qualified as a town planner and worked in local government for a number of years. He now works as a freelance photographer and writer, specialising in environmental issues. He lives in London.

Nuclear Juggernaut

The Transport of Radioactive Materials

Martin Bond

Earthscan Publications Ltd, London

First published 1992 by
Earthscan Publications Ltd
120 Pentonville Road, London N1 9JN

British Library Cataloguing in Publication Data
Bond, Martin
  Nuclear juggernaut: the transport of radioactive materials.
  I. Title
  363.17

ISBN 1–85383–103–4

Typeset by Bookman Ltd, Bristol BS3 1DX

All photographs © Martin Bond with the exception of the following:
Greenpeace/Gleizes (p. 2); P. Cartlidge (p. 38); *Western Morning News*
(p. 40); Martin Jenkinson (p. 41); Greenpeace/Gibbon (p. 78); anon.
(p. 132); *Bristol Evening Post* (p. 137); *Chatham, Rochester and Gillingham
News* (p. 174); Nukewatch (USA) (p. 193).
Front cover: A Unifetch flask for transporting spent fuel from research
reactors, on the A74 north of Beattock, Scotland.
Back cover: A Type A package, containing the radioisotope rubidium-81,
in the guard's van of a Liverpool passenger train, Euston Station, London.

Earthscan Publications Limited is an editorially independent subsidiary
of Kogan Page Limited, publishing in association with the International
Institute for Environment and Development and the World Wide Fund for
Nature (UK).

Contents

Acknowledgements

I would like to thank the following whose assistance in ways large and small helped to make this book and the photographs possible: Simon Bennett; Martin Forward; Ian Leveson; Steve Martin; William Peden; Pete Roche; Greenpeace International (and especially Andy Sterling, Simon Carroll and John Willis); the late, lamented Greater London Council; the staff of the National Radiological Protection Board; and Tom Horlick-Jones and Jenny Edwards of the London Emergency Planning Information Centre. Thanks also to Tarin, Gill and sister Sue for help "on the road"; to Kath for the buckwheat pancakes in Chester without which the front cover photo would not exist; and to my editors Neil Middleton and Sian Mills for their patience and flexible deadlines. Finally, a special acknowledgement to Margaret Savage-Jones and Janice Owens without whom I'd have done something totally different!

Glossary

Activity	The amount of radioactivity in radioactive material. Generally an abbreviation of radioactivity.
Alpha particle	A particle emitted during the decay of some radioactive atoms. It comprises two protons and two neutrons.
Becquerel (Bq)	The SI unit of radioactivity, corresponding to the decay of one radionuclide per second. One becquerel equals 2.7×10^{-11} curies, the unit of radioactivity it has now replaced. (kBq = kilo or 1,000 bequerels; MBq = mega or 1,000,000 bequerels; GBq = giga or 1,000,000,000 bequerels; TBq = tera or 1,000,000,000,000 bequerels).
Beta particle	Electrons or positrons emitted during the decay of some radioactive atoms.
Collective dose	Strictly speaking, the *collective effective dose equivalent* – the quantity obtained by multiplying the average effective dose equivalent by the number of people exposed to a given source of radiation and expressed in man-sieverts (man-Sv).
Contamination	The presence of radioactive material where it is not wanted.
Core	The part of a nuclear reactor which contains the fuel.
Curie (Ci)	The old unit of radioactivity, now replaced by the Becquerel. One curie equals 3.7×10^{10} becquerels.
Decay	The spontaneous transformation of a radionuclide. The decrease in the activity of a radioactive substance.
Depleted uranium	Uranium in which the proportion of the isotope U–235 has been reduced, either below the natural level (0.7 per cent by weight) or below a former level of enrichment.
Dose	General term for the quantity of radiation received by one or more people. (Also used – not strictly accurately – to indicate the quantity of radiation emitted by certain transport packages.)
Dose equivalent	The quantity obtained by multiplying the absorbed dose by a factor to allow for the different effectiveness of the various ionising radiations in harming tissue. Measured in sieverts.
Enrichment	The process of increasing the proportion of one isotope

	relative to another of the same element.
Fast reactor	A reactor in which there is no moderator to slow down neutrons for nuclear fission. The chain reaction is sustained by fast neutrons.
Fission products	The new elements produced in nuclear fuel when atoms of uranium and plutonium are split by nuclear fission.
Fuel element	Either a single fuel pin or rod (in Magnox reactors) or an assembly containing an array of individual fuel pins.
Fuel pin	Part of a fuel assembly containing, depending on the type of reactor, uranium and/or plutonium fuel.
Gamma rays	Highly penetrating electromagnetic radiation emitted by certain radioactive atoms.
Half life	Time required for the rate of radioactive emission from a particular substance to fall to half its original value.
Hex	Uranium hexafluoride (UF6)
Irradiated fuel	Nuclear fuel which has been fissioned inside a reactor and therefore contains fission products. Also known inside and outside the industry as *spent fuel* although this is not strictly accurate.
Isotope	Atoms of a given element with different numbers of neutrons within their nuclei.
Light water reactors	Reactors which use ordinary water as a moderator.
Magnox	An alloy of magnesium used to clad natural uranium fuel for the first generation of British nuclear power station.
Moderator	A substance used in thermal reactors to reduce the energy of neutrons produced in the fission process so that they may cause further fission more readily.
Neutron	A sub-atomic particle of zero net charge, with a mass approximately equal to that of a proton. It is a basic component of the atomic nucleus.
Nuclear waste	Common but inaccurate name for irradiated fuel – only the small proportion of fission products created by a nuclear reaction can properly be described as 'waste'.
Nuclide	A species of atom differentiated by the number of protons and neutrons present.
Package	Generic name for the various containers and flasks, etc. used for the transport of radioactive materials.
Proton	A sub-atomic particle carrying a single positive charge and with a mass approximately equivalent to a nucleus. It is a basic component of the atomic nucleus.
Radiation	Energetic particles (alpha, beta and neutrons) or electro-magnetic waves (gamma rays) given off by atoms when they decay.
Radioactivity	The spontaneous disintegration of certain atomic nuclei.
Radiological	Pertaining to the effects of radiation exposure.
Radionuclide	An unstable nuclide that emits ionising radiation.
Rem	An old unit measuring dose equivalent, now replaced by the seivert. 1 rem = 0.01 Sv.

Reprocessing | The chemical treatment of irradiated fuel to separate unused uranium and plutonium from the fission products.

Roentgen (R) | An obsolete unit for measuring the quantity of radiation, or radiation dose.

Sievert (Sv) | The unit for measuring radiation dose, dose equivalent or effective dose equivalent. (Microsievert = one millionth; millisievert = one thousandth).

Tonne (t or te) | 1,000 kilograms or 0.984207 imperial tons (2240 lbs).

Yellowcake | Common name for uranium ore after it has been processed by a mill.

Introduction

The transport of radioactive materials is totally safe. At least that is the view of the nuclear industry and the government. According to the International Atomic Energy Agency (IAEA) "In more than 40 years of experience, there have been no known deaths or injuries due to the radioactive nature of (the) material being transported. . .".[1] Another reassurance, from an advisory committee set up by the Department of Transport, goes even further: ". . .there has been no known damage to the health of anyone due to the transport of radioactive materials. . .".[2]

And yet despite this record of apparent perfection, the movement of various kinds of radioactive material – from nuclear warheads to nuclear waste – has generated anxiety, criticism and outright opposition, including demonstrations against various targets:

- 1984   French protestors try to prevent a consignment of plutonium sailing to Japan.
- 1985   Uranium exports sailing from Darwin, Australia, are held up by anti-nuclear activists.
- 1988   Liverpool dockers refuse to handle cylinders of uranium hexafluoride (a uranium compound from which nuclear fuel is made) on their way to the United States.
- 1989   CND groups on south Humberside demonstrate at the offices of Exxtor, a shipping company which transported nuclear fuel materials across the North Sea.
- 1990   Swiss demonstrators chain themselves to railway lines at Goesgen nuclear power station, to prevent a flask of irradiated nuclear fuel departing for Sellafield.

To those in the nuclear industry, it must sometimes seem that the mere sight of a radiation symbol on the move is guaranteed to provoke hostility from the public and from environmental groups, local authorities and the press. Travelling around on our roads and railways, in the air and over the sea, are nuclear materials of the sort that vaporised Hiroshima and

Nagasaki or fuelled the fated reactors at Chernobyl and Three Mile Island. As other aspects of the industry have proved controversial, it would be surprising indeed if the movement of its raw materials had not also attracted criticism.

Demonstrators in Dunkerque protest against spent fuel deliveries to Sellafield. These Castor flasks were transporting spent fuel from Germany.

Yet the transport side of the industry is one area where there has not – as far as is publicly known – been a major disaster. The industry will point to the various regulations which control the transport of radioactive materials. The International Atomic Energy Agency (IAEA) provides model regulations, adopted by national governments including Britain, which, for example, recommend approval tests to verify the integrity of the various packages, containers and flasks and specify limits for radiation emissions in transit. Containers used for the most dangerous materials, such as spent fuel, must survive drop tests from a height of 9 metres and an 800°C fire for at least half an hour. In the industry's view, the regulatory framework and its own good practice have ensured an exemplary record.

But if that is the case, why do shipments of materials like uranium and plutonium attract such a fuss? Is it a gut reaction against all things nuclear, or do movements of radioactive materials pose more of a threat than the industry is prepared to admit? Alternatively, are protests against such movements simply a tactical ploy by anti-nuclear campaigners to hinder the activities of the industry in an area where it is vulnerable to disruption?

Undoubtedly, public anxiety about radiation provides one explanation. All radioactive substances are potentially hazardous and must be transported in flasks, containers or some other type of package designed to prevent their contents escaping to the environment: not without reason must they be labelled with radiation warning symbols (with the exception of some items with extremely low levels of radioactivity). Questions of health and safety are inevitable: will radioactivity leak in transit? What would be the consequences of an accident? Look at the history of any other part of the nuclear industry and something has always gone wrong – from the cancers and contamination that have followed uranium mining; to accidents involving every type of reactor; to leaking storage tanks of radioactive waste. Is the *transport* of radioactive materials immune to error, or do accidents simply never happen?

Unfortunately they do, and over the years most types of radioactive material – from uranium ore imports to nuclear warheads – have been involved in transport accidents. So far, most have not been too serious and few have led to a release of radioactive material. From the industry's point of view, the worst damage has probably been the unwanted publicity that tends to ensue. Thus the 1984 sinking in the Channel of the French ship the *Mont Louis*, which had been transporting uranium hexafluoride to the Soviet Union, gave added weight to protests against the shipment from France, just a few weeks later, of a consignment of plutonium, exported to Japan by sea.

Three years later the crash, near the Dean Hill Royal Ordnance Depot in Wiltshire, of a Ministry of Defence lorry used for carrying British nuclear warheads drew public attention to the fact that Britain's nuclear deterrent is not confined to military bases. Nuclear warheads move around the world by road, air and sea and in the event of a major accident their contents could be potentially fatal even without being used in war.

Given the volume of radioactive material moving around, it would be unusual if accidents did not happen occasionally. Some ten million packages containing radioactive materials are transported worldwide every year[3] – with about half a million of those in Britain. Despite reassurances of safety, such statistics suggest other reasons for concern: ten million packages of what, for example? Unfortunately, attempts to find out exactly what *is* moving around show that health and safety issues are not the only concern. In common with many other aspects of the nuclear industry, the transport of radioactive materials is bedevilled by secrecy. Movements are unannounced, local authorities are not informed, and certainly individual citizens, concerned about something passing through their neighbourhood, stand little chance of finding out. The movement of most radioactive materials is unknown to the public at large and the industry is happy to keep it that way.

As with other nuclear activities, it sometimes seems that information is released only when something goes publicly wrong. The *Mont Louis* incident, for example, drew much attention to Europe's nuclear trade with the USSR, including Soviet contracts to process nuclear materials for Britain, a sensitive subject in the Cold War era. Such incidents, like the demonstrations above, often reveal hidden corners of the industry's operations – places like Liverpool and Humberside are not usually found on maps of nuclear Britain.

As the statistics suggest, the use – and therefore the transport – of radioactive materials is a lot more extensive than often realised. They are not only used for the production of nuclear electricity and warheads: most movements of radioactive material consist of radioisotopes, widely employed in industry, research and medicine. Although there are some dangerous exceptions, many contain relatively low amounts of radioactivity.

Within the nuclear industry itself, transport operations are also extensive; the image of the solitary lorry making a rare delivery of fuel to a nuclear power station is but a small part of the picture. Nuclear power stations such as Sizewell and Sellafield and military sites like Aldermaston do not exist in isolation. They require a back-up network of nuclear facilities to process the materials that keep the industry going. The radioactive ingredients of a nuclear power station, or a Trident warhead, or a nuclear submarine reactor, will typically have passed not only through several different factories in Britain but also through others around the world.

At the same time, Britain's nuclear industry has an important international role: while many countries have nuclear reactors, Britain is one of relatively few places where nuclear fuel materials can be manufactured and processed. As a result, radioactive materials are imported and exported for overseas customers as well as for domestic use. Britain has become a major link in the international nuclear chain, although the extent of this trade is not always fully appreciated.

Secrecy provides one reason why. The reluctance to divulge information which afflicts Britain's own nuclear programme extends to the processing work carried out in this country for overseas customers; the prerogative of national security is replaced by "commercial confidentiality". Britain's nuclear trade with the rest of the world adds an extra dimension to the transport of radioactive materials. By increasing the quantity of material moving around, it increases any associated risks. Commercial confidentiality makes it harder to obtain basic information about the quantities involved or the transport arrangements. Legitimate questions about, for example, how local authorities might respond to an accident involving imported radioactive material are difficult to answer. Moreover, the industry's overseas trade, especially with countries like South Africa has given the movement of radioactive materials a broader political dimension.

Clearly, the transport of radioactive materials is not just a health and safety issue. As the following chapters explain, military connections, secrecy and politics – domestic and international – all impinge on what the industry views as an otherwise commonplace activity.

2 Radiation, regulations and container design

Radioactivity is defined as the transformation of one chemical element into another as a result of atomic disintegration. It occurs in a few naturally occurring elements, such as radium and uranium, whose atoms are unstable. Over a period of time their atoms change into atoms of other elements which may also be radioactive and disintegrate. This process of decay is accompanied by the emission of ionising radiation from the disintegrating atoms and continues until a stable, non-radioactive element is eventually formed. All radioactive materials emit varying amounts of one or more of the following four types of radiation:

1 *Alpha particles*. These travel only a few centimetres in air. They are easily stopped and will penetrate neither a sheet of paper nor the thickness of human skin. They are, however, a hazard if they enter the human body through a wound, by ingestion, or especially by inhalation.
2 *Beta particles*. These are more penetrating than alpha particles but can be stopped by relatively thin layers of water, glass or metals like aluminium. They can penetrate a centimetre or so of human tissue and are therefore a relatively superficial external hazard; internal organs are more vulnerable to beta particles absorbed into the body internally.
3 *Gamma radiation*. Unlike alpha and beta particles, these are highly penetrating electromagnetic waves which will pass through the human body. Gamma radiation is therefore dangerous whether the source of emission is inside or outside the body. However, intensity decreases with distance from source. Alternatively, the intensity can be reduced by a substantial thickness of solid material. A ten-fold reduction in gamma radiation can be achieved by approximately:[1]

- 1.75 inches (4.45 cms) of lead
- 3.5 inches (8.89 cms) of steel
- 13 inches (33.02 cms) of ordinary concrete
- 20 inches (50.8 cms) of brickwork

4 *Neutrons (sub-atomic particles).* These are produced spontaneously by the decay of certain fissile materials. Like gamma radiation, neutrons are highly penetrating and radiologically equally dangerous.

IAEA regulations

Radioactive materials are transported in accordance with national and international regulations. These are mostly derived from the International Atomic Energy Agency's (IAEA) "Regulations for the Safe Transport of Radioactive Material" (Safety Series No. 6). The most recent edition appeared in 1985. Their objective is to: ". . . protect the public, transport workers, and property from both the direct and indirect effects of radiation during transport". The potential radiation dangers identified by the Agency are:

1 the dispersion of radioactive material and its possible uptake by people nearby during normal transport or in the event of an accident;
2 the hazard due to radiation emitted from the package; and/or
3 the possibility that a chain reaction (criticality) may be initiated in the material contained in the package.

Protection against these dangers is achieved by ". . . limiting the nature and activity of the radioactive material which may be transported in a package of a given design, specifying design criteria for each type of package, and recommending simple rules for handling and stowage during transport".[2] At the heart of the Agency's regulations are broad specifications for four general types of transport package which between them accommodate the vast range of radioactive materials transported. The regulations also set limits on the amount of radiation that may be emitted from a package and the amount of radioactive material that may be released in normal and accident conditions. To demonstrate the ability of packages to transport radioactive material safely – especially in the event of an accident – the Agency recommends design standards and tests for each type of package. The regulations prescribe four basic types of package. In increasing order of robustness they are:

Excepted packages

These contain quantities of radioactive material sufficiently small to allow exemption from most design and use requirements. Radiation dose rates at the surface must be below 0.005 mSv per hour. There are, however, certain requirements to ensure that they are safely handled and transported.

Industrial packages

These are used for transporting materials defined either as low specific activity (LSA), such as yellowcake and low-level radioactive waste which have low levels of radioactivity per unit mass, or surface contaminated objects (SCO) – non-radioactive objects with low levels of surface contamination. Both types of materials are considered safe because the radioactivity is low or because the material is in a form which cannot easily be dispersed. Conventional metal drums used for yellowcake shipments and ordinary freight containers fall into this category. There are three types of industrial package, graded according to integrity: IP-1, IP-2, and IP-3. IP-1 packages must meet certain temperature and pressure requirements; IP-2 packages must be subjected to drop and stacking tests; and IP-3 packages must satisfy additional water spray and penetration test requirements.

Type A packages

These packages provide an economical method of transporting relatively small quantities of radioactive materials, like radioisotopes. They must be robust enough to survive minor accidents in transit, such as falling from vehicles or being dropped, without releasing their contents. Package designs must meet the requirements of the industrial IP-3 package with additional test requirements if the material to be carried is in liquid or gaseous form.

Because most Type A packages are constructed of lightweight, low-strength components they are not designed to withstand a major transport accident. In such a situation it is assumed that some of their contents might escape. The IAEA regulations therefore limit the amount of radioactive material that can be transported in Type A packages. It is assumed that in an typical severe accident, 0.1 per cent of the contents of a Type A package will escape and that 0.1 per cent of that release (i.e. a millionth of the original contents) will be taken up by a rescue worker or member of the public. Taking the ICRP's maximum permissible body burden for different radionuclides as a guide to their radiotoxicity, the IAEA derived radioactivity contents limits for Type A packages for each different radionuclide.

Type B packages

These transport the most highly radioactive materials, including irradiated nuclear fuel, high level nuclear wastes and the more radioactive radioisotopes. In contrast to Type A packages, Type B designs, such as spent fuel flasks and plutonium containers, are supposed to be able to survive all kinds of severe accidents without releasing any radioactive material. Type B packages must therefore be tested to show their resistance to impact and fire. The number of movements made by each type of package gives a rough indication of the types of materials transported around: 48 per cent of all movements are made using Excepted packages; Type A packages account for 41 per cent, Industrial containers 9 per cent and Type B designs a mere 2 per cent of the total.

IAEA tests

IAEA tests are performed on prototype packages as part of a process of gaining approval from an appropriate national authority for a particular design. Approval is forthcoming only if a package "passes" the tests. IAEA tests for demonstrating package integrity comprise two groups; Type A tests which are also applicable to Type B designs and are intended to simulate "normal" conditions experienced during transport – rough handling, minor mishaps and exposure to rain – and more rigorous Type B tests to demonstrate an ability to withstand severe accident conditions. Type A tests are prescribed by the IAEA as follows:

Water spray test

A package is subjected to a water spray approximately equivalent to a rainfall of 5 cms per hour, uniformly distributed for one hour.

Free drop test

The package is dropped onto a hard target (a concrete block) so that maximum damage is inflicted on the safety features. The lowest point of the package must drop at least 1.2 metres before hitting the target. Packages weighing more than 1.5 tonnes are dropped from a lower height.

Compression test

The package is subjected for 24 hours to a compressive load equal to either five times the weight of the package or the equivalent of 1,300 kg per m^2 multiplied by the vertically projected area of the package, whichever is the greater. The load is applied to the top of the package.

Penetration test

The package is placed on a rigid, flat, horizontal surface, and a 6 kg bar of 3.2 cm diameter is dropped from a height of one metre to fall on the centre of the weakest part of the package.

Further tests are required for Type A packages which are designed to transport liquids and gases. An additional 9-metre drop test is required to inflict maximum damage to the containment and another penetration test must be performed, this time from a height of 1.7 metres.

For Type B packages, additional tests are required to demonstrate their ability to withstand severe accidents. The first two, mechanical and thermal tests, are designed to test the cumulative effects of impact followed by heat and should be carried out in that order.

Mechanical test

A package must be subjected to two successive drop tests to assess the effects of impact and penetration. In the first test, the package is dropped 9 metres onto a flat, horizontal unyielding surface so as to inflict maximum damage. The second test comprises a one-metre drop onto a 15 cm diameter mild steel bar perpendicular to, and projecting at least 20 cm above, the target surface. For some light-weight Type B packages, an alternative to the nine-metre drop test was introduced in the 1985 edition of the regulations. This comprised a crush test, where a package is placed on the unyielding target so that maximum damage is experienced when a 500 kg steel mass is dropped on it from a height of 9 metres.

Thermal test

After the mechanical test, the same package is exposed to a temperature of at least 800°C for not less than 30 minutes.

Water immersion test

Most packages must be immersed in water at a pressure equivalent to a depth of at least 15 metres for not less than eight hours. A further test was included in the 1985 edition of the regulations: some packages designed to transport irradiated fuel must be immersed in water at a pressure equivalent to a depth of at least 200 metres for not less than one hour.

Radiation control and labelling

The IAEA regulations specify radiation dose limits for transport packages, which must be labelled accordingly. In the case of excepted packages (where by definition surface radiation levels must not exceed 0.05 mSv), no label is required on the outside of the package. Other types of package must display a label containing the familiar trefoil radiation warning symbol and an indication of the surface dose rate. This is given in two ways: first, by classifying packages into three categories (I-III) defined by the maximum radiation levels (in mSv/hour) allowed at the surface and at a distance of one metre. The category is indicated on a label by one, two or three red stripes. Second, Category II and III labels include a "transport index" (TI), a figure which equals the maximum radiation level at one metre from the surface of the package. The transport index is measured in the old unit of millirems per hour; thus a label with a TI of 6 indicates a radiation dose at one metre of 0.06 mSv per hour. Radiation levels for the three types of package are delimited as follows:

	Surface	1 metre	Transport Index
Category I (White label)	0.005	0	–
Category II (Yellow label)	0.5	0.01	< 1
Category III (Yellow label)	2.0	0.1	< 10

In the case of Category III, higher limits apply if packages are transported under "exclusive use" (formerly known as "full load"), that is, if the vehicle or freight container or other means of transport is carrying no other cargo. Surface radiation levels can be as high as 10 mSv per hour with no limit to the transport index.

The Category I-III classification does not correspond to any particular type of package; Industrial, Type A and Type B packages may require any of the three labels although in practice radiation levels may be well below the maximum permitted for a particular category.

The transport index determines how many packages can be transported together. If the mode of transport – for example, a freight container – carries no other freight, radioactive materials can be transported under "exclusive use" where, with the exception of fissile materials, no limits are imposed on the transport index total. Transported with other goods or passengers, the transport index total is limited, mostly to 50 (for example, in the case of vehicles or passenger aircraft), but in some situations, such as freight-only aircraft, to 200.

The transport index, or the sum of the transport indices if several packages are transported together, also determines how far radioactive material (other than excepted and category I packages) should be segregated from other freight or people. Segregation distances – for example, for use in aircraft also carrying passengers – are set out in tables according to radiation levels (the transport index total) and the duration of the journey. (The figure is calculated differently for fissile materials, see below.)

Implementation

The IAEA regulations are not themselves mandatory; they are, in effect, "model" regulations which the Agency recommend for adoption by governments and organisations involved in the transport of radioactive materials. In many countries they have been incorporated into national legislation or adopted as advisory codes of practice. They have also been adopted by national and international organisations concerned with transportation, such as British Rail or IATA (International Air Transport Association). With non-governmental organisations, the regulations have no statutory force but constitute "conditions of carriage".

Under the regulations, responsibility for safety is split between three parties: the "competent authority" – the national or international body responsible for implementing the regulations; the "consignor" – the person, company or organisation dispatching the material; and the "carrier" – the person, company or organisation transporting it.

In Britain, the competent authority is the Secretary of State for Transport, in practice the Department of Transport, whose executive functions in this area are carried out by its Radioactive Materials Transport Division. These include ensuring that the IAEA's regulations are translated into British legislation, and issuing approval certificates for the use of different packages. For road transport, the IAEA regulations are effected by the Radioactive Substances (Carriage by Road)(Great Britain) Regulations 1974 (SI 1974 No.1735); for rail transport, by British Rail's Conditions of Acceptance of Dangerous Goods (British Rail publication BR 22426, 1977). For sea and air transport, the respective provisions are contained in the Merchant Shipping (Dangerous Goods) Regulations 1981 (SI 1981 No. 1747) and the Air Navigation Order 1980 (SI 1980 No. 1965).

The process of issuing a certificate ensures that flasks conform to the standards laid down in the IAEA regulations. (In some cases, packages intended for small quantities of radioactive material (including many radioisotopes) or materials with low levels of radioactivity, will not need a certificate.) The applicant, usually a company within the nuclear industry (such as BNFL), must submit a safety case for the package to the Department and carry out appropriate tests to show that the package complies with the IAEA's regulations. The Department of Transport assesses the design and, if satisfied, issues a certificate of approval.

Certificates will detail what, and how much, radioactive material the package can hold. Adhering to such specifications is one of the responsibilities of the consignor who must ensure that an appropriate package is used. The consignor is also responsible for ensuring that the radioactive material is properly packed and labelled and that the documentation is in order.

Criticality

Transport packages for certain radioactive materials must guard against the possibility of a "criticality" accident – an unwanted chain reaction. This can only occur in materials which are "fissile", principally plutonium-239 or uranium-235. If assembled in sufficient quantity and in a certain shape and form, fissile materials are capable of maintaining a self-sustained nuclear chain reaction, the basis of nuclear explosions and nuclear reactors. Criticality can occur by accident, either by bringing sub-critical quantities of material too close, or by introducing a moderator such as water; this slows down stray neutrons making them more likely to cause nuclear fission – the process by which most nuclear reactors operate.

Although an accidental chain reaction would not produce a nuclear explosion, it would nevertheless release gamma rays, neutrons and heat – technically known as a nuclear "excursion". If the wrong circumstances prevailed, the pressure created by a rapid generation of heat might burst a transport container causing – according to the UKAEA – "a small explosion" comparable to an exploding car petrol tank.[3] The accompanying radiation could, if severe enough, be lethal to people nearby.

In practice, criticality accidents should be easily avoidable. The simplest precaution is to restrict the quantity of fissile material to an amount which cannot be made critical: criticality cannot be achieved with less than a certain minimum mass of material. Transport containers must also be impervious to water. Quantities of fissile material larger than a critical mass can be made subcritical by incorporating neutron-absorbing materials into the transport package. The transport packages for fissile materials typically incorporate layers of wood and cadmium which surround an inner container holding the radioactive material – wood slows down stray neutrons, which are "captured" by the cadmium.

These considerations are reflected in the IAEA regulations. Packages of fissile material must come under one of four categories:

1 *Exempt.* Quantities and forms of fissile material which could not be made critical under any transport circumstances.

2 *Fissile Class 1.* Packages which are nuclearly safe in *any number* and in any arrangement during transport. Nuclear safety is ensured by providing neutron-absorbing materials in the packaging thus preventing neutrons from one package interacting with other fissile material.

3 *Fissile Class II.* Packages which in *limited numbers* are nuclearly safe during transport. An appropriate "allowable number" of packages is defined by a formula based on the transport index.

4 *Fissile Class III.* Packages which are nuclearly safe by virtue of *special arrangements.* These may be based on a limited mass or configuration of material, its containment, or on special transport arrangements to keep it away from other fissile material.

3   Yellowcake

When Prime Minister Margaret Thatcher opened the Torness nuclear power station in Scotland in 1989, she declared: "The energy which comes from one delivery of nuclear fuel to Torness by road provides the same amount of heat as 20,000 lorry-loads of coal."[1]

Mrs Thatcher's statistic may well be true, but if it was meant to compare transport operations in the different industries it was more than a little misleading. Coal is a combustible mineral which can be dug straight out of the ground and burned. Uranium, by contrast, must be extensively processed before it can be used as a nuclear fuel. Indeed extensive treatment is required just to extract uranium from its parent rock, as the Prime Minister must have known. During a trip to Namibia in the same year, Mrs Thatcher visited Rio Tinto Zinc's uranium mine at Rossing. Between 1977 and 1984, Rossing supplied some 8,600 tonnes of uranium ore concentrates, otherwise known as yellowcake, to Britain for domestic use. That concentrate was the end product of mining and milling the granitic rocks of the Namib Desert, the first stage of which has been described by RTZ themselves: "The scale of the mining operation at Rossing in full production can be illustrated by the fact that the equivalent of a 100-truck goods train loaded with rock leaves the open pit every 20 minutes, 24 hours a day, 365 days a year."[2] Compare *that* with 20,000 lorry-loads of coal! In fact, at Rossing the transport of ore from the mine to the mill is carried out by a fleet of trucks, an operation so intensive that in 1986 the road network was electrified with a "trolley assist" system.

The percentage of uranium in the rocks of the Namib Desert is small. A typical month at the mine might see over 3 million tonnes of rock removed to supply almost one million tonnes of ore to the mill, to yield just 350 tonnes of yellowcake. The volume of rock that must be mined to extract such a tiny percent of uranium explains why uranium mills are located in the mining areas. However, by the time the rock has been crushed, processed into yellowcake and packed into drums for dispatch, transport requirements are reduced. Typically, the annual fuel needs of a 1,000 megawatt nuclear power station can be met by the contents of seventeen 20-foot freight containers.

The major uranium producing areas of the world are in North America, Africa and Australia. In 1989, total production (excluding that consumed by current and former centrally planned economies) was approximately 40,000 tonnes (U3O8) of which Canada accounted for a third, the United States 15%, Australia 9.8 per cent, France 9.7 per cent, Namibia 8.9 per cent, and Niger and South Africa 8.6 per cent each.[3] By contrast, few of the countries with nuclear power programmes, particularly in Europe and the Far East, have indigenous supplies of uranium. Exceptions include Canada and South Africa – both of which produce far more than they consume – and the United States and France which are partially self-sufficient. The Soviet Union and China are also substantial producers, although mainly self-contained. For most other countries, including Britain, nuclear programmes depend upon imports.

As a result, thousands of tonnes of radioactive ore are transported each year from sources often halfway round the world from the customer, a global trade which has provoked opposition in several countries. Accidents provide one reason for concern. In 1986, for example, a fire broke out in bags of cocoa in the hold of the *Meadowbank*, a British registered ship operated by Bank Line, shortly after it had sailed from Darwin in Northern Australia. The *Meadowbank*'s cargo had also included 34 containers of Australian yellowcake destined, via Rotterdam, for Britain, France, the United States and West Germany. Although none of the containers of uranium were damaged by the fire – apparently a case of spontaneous combustion – the incident led to a request in the Australian parliament for a halt to further uranium exports on the Bank Line service pending a safety assessment of their ships. The request was refused.

By and large, however, the transport of uranium ore has attracted attention not because of the hazards of radioactivity in transit – yellowcake's low level emissions are not usually seen as a major health and safety issue – but because of its strategic status in relation to other nuclear and political issues: uranium is the raw material for both nuclear weapons and nuclear power. It is ultimately the source of radioactive pollution, fallout and waste, not to mention the world's nuclear arsenal. Yellowcake shipments offer a convenient opportunity for protest; a campaigning target for those opposed to the rest of the nuclear industry and its associated problems.

In exporting countries like Canada and Australia, demonstrations against yellowcake shipments have also been part of campaigns against uranium mining. Like the extraction of other minerals from the ground, mining the raw material for nuclear weapons and nuclear energy degrades the environment. As uranium ores are also radioactive, the familiar dangers of mining are augmented: miners are exposed to radioactive gases and dust while problems with radioactive waste disposal and contamination have affected neighbouring communities and the surrounding environment. Thus when the first consignment of uranium from the new Olympic Dam mine in South Australia sailed from Adelaide in November 1989, it was blockaded by demonstrators protesting about contamination and the depletion of water supplies.

Demonstrations against yellowcake shipments have also sought to draw attention to uranium's military potential. Several years earlier, on the other side of Australia, Greenpeace campaigners attempted to blockade yellowcake exports from Darwin. Direct actions aimed to highlight the dangers of nuclear weapons proliferation and the inadequacies of international non-proliferation safeguards. Such concerns have not been confined to environmental groups. From 1984 until 1986, Australia's Labour government banned shipments of Australian uranium to France until that country stopped testing nuclear weapons in the Pacific.

In Britain, attention has focused mainly on uranium imports from South Africa and Namibia. Deliveries from both countries have been affected over the years by international sanctions which Britain has consistently opposed or ignored. Attempts to publicise or protest against the transport of imports from southern Africa have provided a means of publicising Britain's links with apartheid. Uranium imports fell foul of the broader struggle for liberation. Deliveries from Rossing in particular have raised other nuclear issues, drawing attention, for example, to the effects on the workers and communities involved in mining uranium and the potential military use by Britain of the imported product.

Britain's uranium supply

Between 1985 and 1989, Britain's nuclear power stations were consuming an average of 1,740 tonnes of uranium a year. Annual consumption by the year 2000 is expected to decline to around 1,200 tonnes.[4]

At the time of writing (1991), uranium supplies for Britain's nuclear power programme were arriving from Canada, the United States and Australia. The longest running contract has been with Rio Algom in Canada, with uranium supplied from its Panel and Quirke mines in Ontario. However, uranium production at these mines ceased in August 1990 and deliveries to Britain will end in 1992 when the current contract expires. In 1987, imports began from two US companies – Energy Fuels Nuclear (EFN) of Denver, Colorado and the Everest Minerals Corporation of Corpus Christi, Texas. Both companies signed contracts with the CEGB in 1985 to supply 1,588 tonnes of yellowcake over a ten-year period from 1987.[5] The third source of current imports is Australia. An average of 300 tonnes of yellowcake per year arrives from the Olympic Dam mine at Roxby Downs in South Australia. Deliveries have been scheduled from 1989 to 1998.[6]

In addition, Britain's nuclear industry supplements its long-term contracts with the occasional "spot market" purchase. Some can be fairly substantial. In 1978, 1,000 tonnes were reportedly bought from Niger[7] while in 1983 "a few hundred tonnes" were purchased by BNFL from an unspecified source to provide a "float" for its expanding uranium enrichment facilities at Capenhurst.[8]

Other spot market purchases have highlighted British connections with a more controversial source of supply, namely South Africa. Apart from its record of apartheid, South Africa's nuclear weapons capability and its refusal, until 1991, to sign the Non-Proliferation Treaty led to widespread international concern. In 1984, for example, a resolution passed at the IAEA's general conference called on member states – including Britain – to "end all nuclear cooperation with the South African regime . . . and to reconsider their purchases of uranium from South Africa".

Despite this, details were revealed in the following year of two CEGB deals both of which involved South Africa. In the first, the Board purchased a quantity of US-origin uranium from Escom, the South African Electricity Supply Commission. At around the same time, the CEGB also bought a quantity of uranium mined in South Africa itself. This secondhand purchase, completed in October 1984, was reportedly from Spain which no longer needed it.[9] Shortly afterwards, Foreign Office Minister Malcolm Rifkind stated categorically: "let me assure you that we do not import uranium from South Africa".[10]

The gap between recent imports and the somewhat greater annual requirement for new fuel has been filled mostly by uranium recycling. Thus by the end of the 1980s, some 70 per cent of the fuel used in the Advanced Gas-cooled Reactors had been manufactured from recycled uranium (over 15,000 tonnes) recovered at Sellafield over the years by reprocessing fuel from the magnox power stations.[11]

Uranium imports – Britain's international role

The quantity of uranium imported each year into Britain is an official secret: the monthly overseas trade statistics show only aggregated totals of imports of all radioactive materials; neither individual substances nor countries of origin are identified separately. Such secrecy might seem surprising as basic details about supplies for the British nuclear power programme – the countries of origin and the quantities involved – are publicly known. Moreover, the government itself has published figures showing annual uranium consumption by Britain's nuclear power stations. "Strategic reasons" provide one explanation.

Another lies about five miles west of Preston. Off the busy A583 to

Blackpool, a quiet country road leads to BNFL Springfields. Since 1948 it has been the destination of uranium imports to Britain; the place where uranium ores are processed into metal and uranium compounds for civil and military use. Springfields, however, also processes uranium for customers around the world – a major reason why import statistics are sensitive.

BNFL's plant is one of the few facilities in the world which manufacture the feed material for uranium enrichment plants. Enrichment is a process of changing the isotopic composition of uranium. In its natural state, uranium contains only a tiny percentage (0.7) of the fissile isotope U235, the uranium isotope that undergoes nuclear fission. Virtually all the rest comprises non-fissile U238. In order to produce a more efficient fuel for nuclear reactors, most uranium is slightly enriched; that is, the small proportion of naturally-occurring U235 is artificially increased to about 2 or 3 per cent of the total. For other uses, including nuclear warheads and submarine reactors, uranium is enriched to over 90 per cent U235. There are several methods of enriching uranium but most use a uranium compound which can be turned into a gas – uranium hexafluoride (UF6), commonly known as "hex".

The process of turning yellowcake into hex is called "conversion". After the mining and milling of uranium ore, hex conversion is the next stage in the production of enriched uranium fuels. A few statistics illustrate its international importance: in 1991, over 85 per cent of all power reactors in commercial operation in the world – 357 out of a total of 413 – required enriched uranium fuel. Measured by total generating capacity, a better indication of uranium consumption, the proportion was even higher at 95 per cent.[12] Most of the uranium needed for the world's civil nuclear power industry, and for some military purposes as well, must be enriched – and that requires conversion to hex.

While many countries have nuclear power programmes, only a handful have enrichment or conversion plants. The location of these processing facilities therefore has a crucial and dramatic effect on the transport of radioactive materials, and especially on the worldwide movement of yellowcake. Large-scale hex conversion is carried out by only five plants in the world (outside the centrally planned economies); see Table 3.1.

In practice, yellowcake is usually delivered not direct to the customer but first to one of these conversion plants. As most of the uranium used by the nuclear industry passes through these five factories – in just four countries – there is a geographical concentration at the front end of the nuclear fuel cycle.

Springfields has the capacity to convert 9,500 tonnes of uranium per year although, like the other conversion plants, it operates below full capacity. Annual throughput has recently been some 5,000 tonnes of uranium per year, the equivalent of almost 6,000 tonnes of yellowcake. By comparison, Britain's domestic conversion requirements are relatively modest averaging 1,873 tonnes of uranium per year between 1990 and 1995.[13]

Table 3.1: Locations of large-scale hex conversion plants

Operator	Maximum capacity (tonnes U/year)	Location
Eldorado	14,500 (includes 9,000 MTU of new capacity)	Port Hope, Ontario, Canada
Allied Chemicals	12,700	Metropolis, Illinois, USA
Comurhex	12,000	Malvesi and Pierrelatte, France
British Nuclear Fuels	9,500	Springfields, Preston, UK
Sequoyah Fuels Corp. (a subsidiary of Kerr-McGee)	9,100	Gore, Oklahoma, USA

The difference is BNFL's foreign trade. In 1989, the company claimed about 15 per cent of the world's conversion market – thus 15 per cent of the uranium needed for the world's enriched uranium-fuelled reactors converges on Springfields. In a rare burst of *glasnost*, the quantity involved was given in a parliamentary written answer in 1989. In marked contrast to the secrecy surrounding annual uranium imports, the government revealed that since 1979, while over 10,000 tonnes of uranium had been imported for British use (from Canada, Namibia, and the United States), BNFL had also received over 30,000 tonnes of ore concentrates for processing for overseas customers.[14]

As to where this material has arrived from, commercial confidentiality prevails. Nevertheless, BNFL themselves revealed their sources to an industry conference in 1979. Since the earliest deliveries of pitchblende from the Belgian Congo (now Zaire), Springfields has processed yellowcake not only from the well-known suppliers – Canada, Australia, Namibia, South Africa and the US – but also from producers in Argentina, Belgium, Niger, Portugal, Spain, Sweden, West Germany and Yugoslavia.[15]

Uranium imports and sanctions

Some sources have been more controversial than others – in particular, South Africa and Namibia. South Africa's record of apartheid, its military nuclear ambitions, and its occupation, until 1990, of neighbouring Namibia in defiance of United Nations resolutions, resulted in international sanctions against uranium exports from both countries – sanctions which Britain ignored.

In the case of Namibia, South Africa's mandate to govern the country on behalf of the British Empire (granted by the League of Nations in 1920) had been terminated by the UN in 1966. When South Africa refused to leave, the administration of Namibia's affairs was nominally transferred to the United Nations Council for Namibia whose actions to preserve the country's resources for post-independence included "Decree No. 1 for the Protection of the Natural Resources of Namibia" approved by the Council in 1974. Paragraph 1 stipulated:

> No person or entity, whether a body corporate or unincorporated, may search for, prospect for, take, extract, mine, process, refine, use, sell, export, or distribute any natural resource, whether animal or mineral, situated or found to be situated within the territorial limits of Namibia without the consent and permission of the United Nations Council for Namibia . . .

Its relevance to the transport of uranium was spelt out dramatically in paragraph 5:

> Any vehicle, ship or container found to be carrying animal, mineral or other natural resources produced in or emanating from the Territory of Namibia shall also be subject to seizure and forfeiture by or on behalf of the United Nations Council for Namibia . . .

Four years earlier, the UKAEA had signed a contract to obtain uranium from Namibia. The contract was placed with Rossing Uranium Ltd, whose major shareholder has been the British-based multi-national Rio Tinto Zinc (RTZ). However, despite the subsequent endorsement of Decree No. 1 by the UN General Assembly and the International Court of Justice, it was ignored by the British government, the nuclear industry and RTZ. Supplies for Britain were delivered from 1977 to 1984; imports continued after 1984 to be processed by BNFL for their overseas customers.

A new dimension was added in 1987 when several countries, most notably the United States, introduced their own sanctions against Namibian uranium and against imports from South Africa itself. The US restrictions were part of the Comprehensive Anti-Apartheid Act (Public Law 99-440) which came into force on 1 January 1987. The major victims of this legislation included the two US conversion plants, approximately 20 per cent of whose throughput had been coming from southern Africa. The Anti-Apartheid Act meant that none of this material – being converted for European and Far Eastern customers as well as US utilities – could enter the United States. US legislation was paralleled by similar restrictions in Canada.

The British government's refusal to introduce similar sanctions gave BNFL a market advantage: Springfields and its French counterpart became the only facilities in the western world that could process uranium from southern Africa. The significance of this was enhanced by a loophole in the US legislation – the 1987 Act applied only to yellowcake. Uranium from southern Africa could still enter the US in the form of hex, an exemption which dovetailed neatly with BNFL's business at Springfields. By converting southern African yellowcake to hex, BNFL could help their customers to circumvent US law; an attractive option for both US reactor operators with contracts for South African or Namibian uranium and utilities (in the US or elsewhere) who had enrichment contracts in the States.

The US sanctions highlighted Britain's international role in processing uranium: even though British contracts to buy uranium from Namibia and South Africa had ceased – in 1984 and 1973 respectively – uranium from those countries continued to arrive in Britain for BNFL's overseas customers. Utilities such as Enusa in Spain, Synatom in Belgium, Chubu Electric and Tokyo Electric in Japan and the US company Exxon sent uranium from southern Africa to Springfields for conversion. An indication of the quantities involved was given in 1987 when BNFL themselves acknowledged that "up to half to two thirds" of Springfields' throughput – the equivalent of some 3,000 to 4,000 tonnes of yellowcake each year – came from either Namibia or South Africa.[16] South Africa especially has been a major source of uranium for BNFL's overseas customers. BNFL has a "long-standing agreement" with South Africa's Nuclear Fuels Corporation (Nufcor) – renewed in 1985 – to process South African uranium.[17]

Thus the transport of Namibian uranium became a target for anti-nuclear and anti-apartheid campaigns Organisations like the British Campaign Against the Namibian Uranium Contracts (CANUC), sought to publicise deliveries of uranium making yellowcake transport a sensitive operation. In August 1986, for example, dockworkers in Liverpool refused to handle a consignment of enriched uranium hexafluoride suspected of being made from Namibian uranium. Two years later, the national union imposed a complete handling ban on Namibian minerals in UK ports. Anti-apartheid campaigners emphasised the vulnerability of uranium deliveries by asserting that "an effective blockade in this country remains the first priority".[18]

Finally, on 21 March 1990, Namibia gained its independence. (There was one significant exception: South Africa has retained Walvis Bay, Namibia's only deep water port, from which Rossing's output is shipped.) Accordingly sanctions against imports from Namibia were lifted. In the following year sanctions against South Africa were lifted after the country signed the Non-Proliferation Act.

The Namibian controversy and the military demand for uranium

If the politics of southern Africa were remote to the average electricity consumer, another aspect of the Namibian deal brought the subject closer to home. In the year after Namibia's first free elections, the British government all but conceded that Rossing's uranium could be put to military use – one of the reasons why Namibian imports had attracted critical attention.

Britain needs uranium both for nuclear weapons and for the reactors of its nuclear submarines. However, buying yellowcake for military use is not as easy as it used to be. Over the years, Britain's traditional suppliers have imposed conditions – "safeguards" – on their uranium exports, stipulating that they be used only for peaceful purposes. Agreements signed with Canada in 1959 and with Australia in 1981 both prohibit the use of supplied material for military purposes or for the manufacture of any nuclear explosive device. What, then, are the alternatives?

One is that the government does not need to buy new supplies: it can draw on an existing stockpile. A reliable estimate, published in 1981, estimated "government stocks" as 8,000 tonnes of natural uranium and 1,000 tonnes of enriched material.[19] At least some of the stockpile is not subject to safeguards – Calder Hall and Chapelcross are fuelled with unsafeguarded stocks purchased before the 1970s – and is potentially available for military use.[20] However, stockpiles do not last for ever and uranium may still be required for defence purposes. With restrictions on the use of Canadian and Australian supplies, it has long been suspected that imports from Namibia might be put to military use.

One reason was the existence of a secret contract, signed by Prime Minister James Callaghan's Labour government in 1976, to buy a further 1,100 tonnes from Rossing. This was in addition to the 7,500 tonnes of the original, publicly acknowledged contract. Although successive governments refused to confirm its existence – heightening suspicion that it had a military significance – it was later recalled by members of Callaghan's cabinet. Barbara Castle recorded the matter in her political diaries, published in 1980. Castle noted that approval to purchase more uranium was given after strong pressure from the Overseas Policy and Development (OPD) committee, dominated by the military chiefs of staff.[21] Tony Benn recalled the contract in 1982, and concluded that "the only real interest in Rossing must have been for the weapons programme".[22] Castle's diary comments on the subject were belatedly raised in the House of Lords in 1988, leading the government to admit that a contract had indeed been signed for an extra 1,100 tonnes.[23] More significantly, the government confirmed that the extra material, despite being acquired "entirely for civil use" and covered by safeguards, could nevertheless be withdrawn from safeguards – thus tacitly acknowledging its military potential.[24]

Safeguards are the crux of the matter. At the Sizewell Inquiry, for example, BNFL stressed that Namibian imports were covered by Euratom safeguards. These were accepted by Britain in 1973 on joining the European Community. BNFL's assertion – in response to CND questions about possible military use – was presumably meant to imply that uranium covered by Euratom safeguards could only be used for civil purposes. But as BNFL must have known, that is not necessarily so. Uranium imported under Euratom safeguards is assigned a code by Euratom to denote its potential use; a code which is determined by the supplier country: if the supplier imposes "peaceful-use" restrictions, the uranium is labelled "P"; if it is sold with no such restrictions, Euratom labels it "N". As the British government has made clear, "N" labelled uranium can be withdrawn from safeguards provided advance notice is given to Euratom.[25] Thus uranium can be imported under Euratom safeguards and still be available for military use. In the following year, Energy Secretary Cecil Parkinson confirmed that all of the uranium from Namibia – a total of 8,600 tonnes-had been labelled "N".[26]

In practice, therefore, uranium does not need to be purchased specifically for defence purposes: it can be bought for peaceful use and then transferred from safeguards after its arrival in Britain, if necessary by being swapped with safeguarded stocks. In this way, all recent imports from southern Africa – including at least one spot market purchase of South African material – must be assumed to have a military potential.

Transport hazards and transport arrangements

Yellowcake (U308) is transported in cylindrical steel drums inside standard freight containers. Imports arrive by sea. Canadian deliveries from Montreal are received at Felixstowe, where uranium also arrives from US ports such as Houston, Savannah and New Orleans. Australian imports are shipped through Felixstowe, Hull and Southampton.

South African and Namibian deliveries arrive mostly at Southampton on regular container services from Durban and Cape Town respectively. British Rail's Freightliner service carries imports to container depots in the north of England – Liverpool, Leeds, and Manchester (Barton Dock Road) have all been used in the past – and road travel forms the rest of the journey. Reflecting the political sensitivity of southern African supplies, other less straightforward routes have also been used. During the 1980s, containers arrived on lorries from Zeebrugge via cross-Channel ferries to Dover and Chatham.

Yellowcake is a chemically toxic, low-level radioactive hazard. It comprises three alpha-emitting uranium isotopes, naturally occurring in the following proportions: U238 – 99.276 per cent; U235 – 0.718 per cent; and U234 – 0.0056 per cent. Yellowcake also emits small amounts of beta and gamma radiation, given off by residual amounts of uranium "decay products'. Alpha and beta particles are stopped by the ore itself, and would penetrate neither the walls of a steel drum nor a freight container. Even if spilled, alpha radiation travels only a few centimetres in air. It is, however, a greater hazard if ingested or inhaled – a hazard that might affect people if exposed to a dusty spillage.

The 23.47 SX Freightliner from Southampton approaches Barton Dock Road container depot, Trafford Park, Manchester with three containers of yellowcake bound for BNFL Springfields. The rest of the journey is made by road.

A container of Canadian yellowcake arrives at BNFL Springfields.

4   Uranium Hexafluoride

Hex is a verb meaning to practise witchcraft. As a noun, it is also the common name for uranium hexafluoride, the raw material for uranium enrichment plants. As part of an industry that has cast an alchemical spell on the second half of the twentieth century, its double meaning is not inappropriate.

The large-scale production of hex was first stimulated by the Manahattan Project, the wartime collaboration between Britain, Canada and the US which led to the atomic bombing of Japan. Hex was the uranium compound used in the process to create the highly-enriched uranium for the Hiroshima bomb. As the inhabitants of this Japanese city were vaporised on 6 August, 1945, they probably weren't aware that it was that very ability to change from a solid to a gas that guaranteed hex such an important place in the nuclear fuel cycle – just as it guaranteed their instant death. Those who were left alive, if not blinded by the ironies of science, may well have thought it was witchcraft.

Hex transport is a product of uranium enrichment. Changing the isotopic composition of uranium also results in different types of hex moving around. Yellowcake conversion produces "unenriched" hex, chemically different from natural uranium but with the same isotopic composition. This is the material delivered to enrichment plants. What comes out is chemically the same but with a different isotopic composition: "enriched" hex (the product stream) and "depleted" hex (the tails stream). All three types are transported in large cylinders.

Although an essential product of the nuclear fuel cycle – for both civil and military use – hex received little public attention until 25 August 1984, when a French ship, the *Mont Louis*, sank in the Belgian shipping lanes off Zeebrugge. On board were 30 full cylinders of hex and 22 empties, on their way from Le Havre to the Latvian port of Riga. As the ship and its radioactive cargo floundered in the sea, uranium hexafluoride floated into the headlines.

Although it has only a low level of radioactivity, hex is chemically toxic and extremely corrosive: in the words of BNFL, it "reacts vigorously with water . . .". In the event of a cylinder being cracked or punctured, BNFL's advice to their drivers is emphatic: "DO NOT USE WATER". Not without reason did the possibility of an explosive underwater radioactive leak dominate many reports of the accident, as did the media's ignorance of the substance involved. Some newspapers, unfamiliar with the subtleties of the nuclear fuel cycle, even called it nuclear waste. The sensationalist tendancies of the media, and a wave of criticism about the sea transport of nuclear materials, prompted BNFL's Chief Executive at the time, Con Allday, to observe: "The extraordinary reaction of the media, some official bodies and at least one trade union to this event appears to me to be an instance of Pavlov's conditioned reflex phenomenon, with neurosis triggered by the word radioactivity".[1]

This would not have been an unreasonable observation, except that the nuclear industry itself is not always the most reliable or consistent source of information. For example, while BNFL's advice to its drivers notes that hex and water react "vigorously", the Health Physics Manager at Springfields asserted at the time that "it did not react violently with water . . ."[2]

Even the UKAEA's own magazine *ATOM* appeared to get it wrong. The November 1984 issue carried an article entitled "Mont Louis – the facts", in which *ATOM* weighed in against press inaccuracy and criticised environmental groups for exaggerating the dangers. They also stated, quite unambiguously, that "All the containers have now been recovered . . ."[3] How surprising, then, that in February of the following year, a hex cylinder was washed up on the beach at Trimingham in East Anglia.[4] Five months later another came ashore on the west coast of Denmark.[5] Like the one recovered in Britain, it was fortunately found to be empty.

The washed up containers, recovered long after *ATOM*'s article, suggested two possible explanations: either they were not from the *Mont Louis* at all, or the nuclear industry's assurances are not always entirely true. Many would suspect the latter and, indeed, the Ringkoebing cylinder was subsequently confirmed as the last to be recovered from the *Mont Louis*.[6] While empty containers are obviously less hazardous than full ones, the incident might be seen as a classic demonstration of the industry's habit of publicly denying that potential problems exist, or that if they do, they can't possibly cause any harm.

Civil hex

Most of the hex transported around Britain is unenriched material produced by BNFL for civil purposes. Over half of their total output has been manufactured for overseas customers, the product of the company's conversion contracts described in the previous chapter. The general pattern of hex movement is determined by the locations of three different types of uranium processing plant, which are linked as follows:

1 Conversion plant
(processes uranium ore concentrates,
"yellowcake", into hex)

Unenriched hex delivered to:
↓
2 Enrichment plant
(produces enriched hex)

Enriched hex delivered to:
↓
3 Fuel fabrication plant
(reconverts enriched hex into a form, usually uranium dioxide, from which fuel elements can be made; or makes the complete fuel assembly as well)

In Britain, conversion and uranium fuel fabrication are both carried out by BNFL at Springfields while enrichment is carried out at Capenhurst by Urenco, a subsidiary of BNFL and Dutch and West German interests. The two sites are linked by a two-way traffic of hex. At Springfields, supplies of unenriched hex go out while enriched material comes in, to be made into fuel or fuel materials. At Capenhurst the pattern is reversed: unenriched hex goes in, enriched material comes out.

In reality, the overall picture of hex movements in Britain is far more complicated because both BNFL and Urenco have contracts to process uranium for overseas customers. At the same time, Britain's nuclear generating companies obtain some of their fuel cycle requirements abroad. These contracts, involving conversion, enrichment and fuel fabrication, generate a complex network of hex imports and exports.

Conversion contracts

BNFL's conversion contracts produce a steady output of unenriched hex from Springfields. While some goes to Capenhurst, much is exported with deliveries to all of the other major enrichment plants in the world. These are run by just four operators (see Table 4.1).

While the United States is a well established destination for BNFL's hex, most of Springfields' output is now delivered to Urenco and the Soviet Union. Shipments to the USSR, for example, include uranium owned by the British generating companies who since 1975 have held enrichment contracts with Techsnabexport. The first covered the supply of enriched uranium over a 10-year period from 1980.[7] In 1985, the CEGB signed a further long-term contract with Techsnabexport for the re-enrichment of 0.4 per cent depleted uranium (recovered from the reprocessing of magnox fuel) to the levels required for AGR fuel.[8]

Table 4.1: Major world civil enrichment plants

Operator	Location
US Department of Energy	Paducah, Kentucky
	Portsmouth, Ohio
	Oak Ridge, Tennessee (now closed)
Eurodif	Tricastin, France
Techsnabexport	Zholtye Vody and/or Chkalovsk, USSR[9]
Urenco	Capenhurst, Britain
	Almelo, Netherlands
	Gronau, Germany

A Type 48Y cylinder for transporting unenriched or depleted uranium hexafluoride, Le Havre, France. (Hex is transported by rail in France but in Britain travels by road.)

In addition, there are shipments between Springfields and the Soviet Union of material belonging to overseas customers. Like deliveries for British utilities, this is unenriched hex converted by BNFL at Springfields for enrichment by Techsnabexport. Traffic to the Soviet Union has been increasing in recent years: in 1988, the government revealed that the total amount of unenriched hex sent from Britain in the previous five years contained some 3,000 tonnes of uranium.[10]

Enrichment contracts

In addition to Urenco's enrichment contracts for the British nuclear power stations, the consortium also enriches uranium for overseas customers. Some of this will have been previously converted by BNFL at Springfields, but other material arrives from conversion plants overseas. Most of Urenco's overseas contracts have been with utilities in the other two countries of the consortium, especially Germany. Other enrichment contracts exist with customers in Sweden, Switzerland and the United States.

After enrichment, Capenhurst's output is delivered to the next stage of the nuclear fuel cycle. In some cases this will be to Springfields for reconversion to uranium dioxide from which nuclear fuel is made. Alternatively, low enriched hex is exported for further processing elsewhere. Companies in Europe which fabricate enriched fuel include Reaktor-Brennelement Union (RBU) at Hanau in Germany and FBFC at Romans in France.

Fuel fabrication and fuel materials

Low enriched hex received at Springfields is processed into uranium dioxide or manufactured into complete fuel assemblies. Much of this work is for the British AGR power stations and Sizewell B. Enriched uranium fuel is also manufactured for export – for example, for reactors in the Netherlands.

Supplies of low-enriched hex arrive at Springfields from other enrichment plants as well as Capenhurst, including Urenco's facilities at Almelo and Gronau. (About half the output from Urenco's three plants returns to Springfields for further processing, while the rest goes to other fuel fabricators.)

Springfields also manufactures uranium dioxide for export. Contracts include the supply of UO2 powder to fuel fabricators in Germany and Spain. The Spanish contract involves selling 100 tonnes of UO2 powder a year to Enusa.[11] Other contracts include UO2 powder for French and Italian utilities, and UO2 fuel pellets for Japan.

The military use of hex

The industrial-scale production of hex in Britain originally began entirely for military purposes, as part of a fuel cycle dedicated to the production of fissile materials for warheads. For many years hex production at Springfields has also been the first stage in the production of highly enriched uranium for fuelling the reactors of British nuclear submarines. Since the the 1960s, the high enrichment stage of this process has been carried out in the United States – under a 1969 amendment to the 1959 Mutual Defence Agreement, Britain supplied the necessary quantity of hex feed to the US in return for highly enriched product.

This arrangement changed in 1986, when BNFL opened the first part of the new A3 enrichment plant at Capenhurst. Originally intended for domestic production of highly enriched uranium for the Ministry of Defence, the project was amended in 1983 to produce uranium enriched to "intermediate level"; high enrichment was to continue to be carried out in the United States as before. Although the exact enrichment level of the plant's output has never been disclosed, BNFL have subsequently described it as "low level", which normally refers to enrichment up to 5 per cent.[12] Thus the military output of hex from Capenhurst is probably similar to the civil product of the Urenco plant next door, that is, around 2 or 3 per cent U235.

The new plant has resulted in changes in hex transport routes: instead of being dispatched directly from Springfields for export, military supplies now go first to Capenhurst and then to the United States. Accordingly, by 1987 when the A3 plant had become fully operational, "government to government" consignments of hex were passing through Liverpool docks.

In due course, however, arrangements will change yet again. In 1991, the MoD announced that it was cancelling its enrichment contract with BNFL.[13] Deliveries will cease in 1993 after which military supplies of hex will, as before, be exported direct to the States.

Hex hazards

Hex moves by all modes of transport: in Britain, it travels entirely by road, while in France deliveries are mostly by rail. International shipments are made by sea and in Sweden, enriched hex has been imported by air. As the quantity of hex moving around has increased, so has concern about its dangers.

At normal room temperature and pressure, hex is a white crystalline solid which gradually evaporates giving off white fumes. Like dry ice (carbon dioxide), it changes directly from a solid to a gas without going through a liquid phase. It is transported as a solid, although when heated under pressure it changes from a solid to a liquid. Hex reacts violently in contact with water and organic compounds including oils and lubricants. A spillage would lead to a rapid chemical reaction with moisture in the air as escaping hex decomposed to form a cloud of uranyl fluoride (UO2 F2) and hydrogen fluoride (HF). The fumes of uranyl fluoride – minute particles, heavier than air – would eventually settle on the ground and be dissolved in atmospheric water. Hydrogen fluoride dissolves in atmospheric water to form hydrofluoric acid. The exact production and distribution of these various substances would depend on the humidity of the atmosphere and the amount of air movement. Although the radioactivity of uranyl fluoride is relatively low, it is one of the most toxic of uranium compounds and potentially lethal.

That potential was tragically demonstrated on 4 January 1986, when an overloaded hex cylinder burst at Kerr-McGee's Sequoyah Fuels conversion plant in Oklahoma. The cause was an attempt, in violation of plant procedures, to empty hex from the cylinder by heating and liquifying the solid contents under pressure. The result was a ruptured cylinder, a 40-minute leak of vapour and, four hours later, the death from acid fume inhalation of plant worker James Harrison, aged 25. At least 32 other people were injured, mostly suffering lung damage, and about 100 people – residents living downwind from the site as well as plant workers – were temporarily hospitalised.

The death of James Harrison was not the first demonstration of hex's fatal potential. Deaths occurred even in the earliest days of the nuclear industry. The NRPB have described a hex leak reported in the 1940s, which killed two laboratory workers and injured several more: "The acute effects of the exposure consisted of corrosive damage to skin, eyes and respiratory mucosa, which was probably caused by the HF (hydrogen fluoride), and transient kidney damage related to the inhalation of uranium".[14]

Normally, hex cylinders arrive at their destination intact. Although accidents in transit have occurred, so far they have not led to leaks. One such incident happened on the Winchester by-pass in Hampshire in 1976. An articulated lorry carrying a full cylinder of unenriched hex inside a shipping container was "nudged" while overtaking another truck. The lorry, on its way from Springfields to the US via Southampton docks, careered across the central reservation and overturned, colliding with three other vehicles travelling in the opposite direction. Police closed a mile-long section of the A33 for three hours while fire officials and radiation experts attended the accident. Despite the full-scale turnout of emergency services, BNFL claimed at the time that ". . . even if the container had been damaged it would not have leaked immediately. And if it did, there would not be any hazard anyway".[15] No mention of possible kidney damage or corroded eyes, nor of the need – to quote BNFL's own internal advice – for protective devices including "self contained breathing apparatus with full-faced mask".

Things might have been different if the Winchester lorry had been caught in a fire. As BNFL have ackowledged, fire presents an added potential danger. Heated within a sealed container, uranium hexafluoride melts under increasing pressure turning into a liquid. If sufficient pressure builds up, a cylinder will eventually explode. According to the Sierra Club (a US environmental organisation which has monitored hex shipments in the States), an accident involving a fire could be lethal out to a distance of three miles, mainly due to the effects of dispersing hydrofluoric acid.[16]

Some idea of the likelihood of a leak was provided in a 1978 report by the United States Pacific Nuclear Laboratories (PNL). As part of an assessment of the risk of transporting hex by truck, PNL estimated the time it would take for a cylinder to fail in an 800 degree C fire. Cylinders used in Britain to transport unenriched hex, the type 48Y, would fail after 48 minutes; type 30B cylinders, used for enriched hex, would fail after only 29 minutes.[17] Not without reason does BNFL's advice stipulate: "keep cylinders cool by spraying with water if exposed to fire but do not spray directly at a point of leakage".

A consignment of type 30B cylinders of uranium hexafluoride (encased in protective overpacks) on the A583 near Preston, bound for the Soviet Union.

The consequences of spraying water on a leak were illustrated in France, just two months after the Winchester crash. A vehicle at the Pierrelatte conversion plant accidentally ripped a tap off a hex cylinder. Over 7 tonnes of liquid UF6 leaked into the atmosphere, turning to gas as it escaped. The escaping hex reacted not only with moisture in the air, but also with water jets sprayed at the leak by emergency workers. The result, according to reports, was a cloud of hydrofluoric acid 3 kilometres long, 2 kilometres wide, and 60 metres thick.[18] Fortunately no one was injured.

The transport of hex has generally been problem free, however. Thousands of deliveries have been made over the years without incident or public concern. Until, that is, the *Mont Louis*. While the British nuclear industry played down the dangers, the ensuing post-mortem raised legitimate criticisms of transport arrangments, not least from within the industry itself. The Commissariat à l'Energie Atomique (CEA) – the French equivalent of the UKAEA – suggested that hex cylinders should be subject to more severe fire tests than at present: not for the first time was the IAEA criterion of 800°C. criticised as too low.[19]

Smit Tak International, who salvaged hex cylinders from the *Mont Louis*, warned a 1986 industry conference of the importance of protecting the cylinders from damage and in particular the valves which, as the Pierrelatte accident suggested, were a potential weak point of the cylinder. Smit Tak itself decided to design new protective covers for the valves and urged an improvement in design standards. Hex manufacturers (such as Comurhex and BNFL) were encouraged to make more information available about cylinder construction strength and to paint them in more distinctive colours than grey.[20]

Hex cylinders are usually covered during transport or containerised. The outline of this load indicates a Type 48Y cylinder.

The vulnerablity of hex cylinder valves again became apparent in July 1987 when the US Department of Energy, which owns the US enrichment plants, discovered cracks in the threads of the valves fitted to 30-inch and 48-inch diameter cylinders. Failure of a defective valve could have caused a leak. The US Nuclear Regulatory Commission ordered the immediate removal of 4,500 suspect cylinders from service.[21]

Trade routes

Trade routes, like the quantities moving around, inevitably vary over time, not least because of BNFL's changing order book – old contracts expire, new ones are won. Transport in Britain is always by road, while imports and exports are by sea. Deliveries between Britain and the USSR are shipped via Ellesmere Port and Riga, courtesy of the Latvian Shipping Company. Deliveries to Urenco's plants in Germany and the Netherlands travel via the M62 to Immingham for Tor Line freight-only ferries to Rotterdam. Hex deliveries between Britain and the US have used Liverpool and Felixstowe docks. French imports arrive at Newhaven on ferries from Dieppe. Other European traffic passes through Dover.

Box 4.1: Uranium hexafluoride cylinders

The three different types of hex – unenriched, enriched and depleted – are transported in two different types of cylinders.

Unenriched and depleted hex is moved in type 48Y cylinders weighing 2.5 tonnes each. Looking not unlike giant kegs of beer with distinctive "hoops", they are 12.5 feet long and 4 feet in diameter. Each hold up to 12.2 tonnes of uranium hexafluoride. The cylinders are made of 16 mm thick steel and pressure tested to 28 bars.

By contrast, low-enriched hex is transported in smaller type 30B cylinders measuring 7 feet long by 30 inches in diameter. Weighing 650 kgs. they have a capacity of up to 2.2 tonnes of product. Low-enriched hex cylinders are covered with a cylindrical 8 feet long by 4 feet diameter "outer protective package". Known as an "overpack", it weighs 820 kgs and provides an extra layer of steel protection.

---

5   Highly enriched uranium

It doesn't quite possess the fearsome aura that surrounds plutonium, but highly-enriched uranium (HEU), like its companion in arms, is a vital ingredient in nuclear weapons. HEU consists mostly of the isotope U235. This occurs naturally in far smaller amounts, comprising only 0.7 per cent of natural uranium, the remainder being mostly U238. If the proportion of U235 is sufficiently increased by enrichment, HEU can be made to form a critical mass and, like plutonium, produce an explosive chain reaction – first demonstrated in 1945 when 60 kgs of HEU incinerated the inhabitants of Hiroshima. Apart from its continued use in weapons, HEU is also manufactured into reactor fuel, principally for the power plants of nuclear submarines. In Britain, there are three distinct uses for the material:

1 In nuclear weapons: HEU, like plutonium, is a fissile material used in British nuclear warheads.
2 In naval reactors: fuel elements made from HEU power the Royal Navy's nuclear submarines.
3 In research reactors: many are fuelled with fuel elements manufactured from HEU.

Used in nuclear warheads and naval reactors, HEU is enriched to over 90 per cent U235. For research reactors, enrichment is typically around 70-80 per cent. Highly enriched uranium is transported around Britain in several different forms: as uranium metal, as new fuel for research and naval reactors, and as irradiated fuel from the same. Because of its military significance, the processing and transport of highly enriched uranium – even for civil consumption – is carried out in considerable secrecy.

From "cradle to grave" – Rolls-Royce and the nuclear submarine programme

Since the launch of Britain's first nuclear submarine – *HMS Dreadnought* –

in 1960, nuclear reactors have powered two types of Royal Navy boat, the ballistic missile submarines – Polaris – and the Fleet submarines, known as "hunter-killers". Although *Dreadnought* was the first to be built in Britain, it was a one-off powered by an imported US reactor, obtained as a result of the 1958 Mutual Defence Agreement between Britain and the United States. The 1958 agreement also resulted in the creation of Rolls-Royce and Associates (RR&A), formed from the consortium which built the first British submarine reactor, the land-based prototype HMS Vulcan next to the Dounreay site in Scotland. RR&A have since designed and supplied the reactors for all of Britain's nuclear submarines since *Dreadnought*. These include the four Polaris boats, *Resolution, Repulse, Renown* and *Revenge*, which will be replaced from the mid-1990s onwards by the Trident submarines. The first, *HMS Vanguard*, was ordered from Vickers Shipbuilders and Engineering Ltd in 1986. When *HMS Triumph* is commissioned – the last of eight Trafalgar class "hunter-killers", the total number of Rolls-Royce powered Fleet submarines in service (or laid up) will have reached 19.

As the size of the fleet has grown, so has the demand for highly enriched uranium fuel. In 1990, however, a report by the Select Committee on Defence – "Options for Change" – considered the implications of the upheavals in Eastern Europe and the Soviet Union and ways in which British forces might be restructured in response. One of the proposals announced subsequently by the government included a reduction in the future size of the "hunter-killer" fleet to 12.[1] Nevertheless, with a new generation of "hunter-killers" in the planning stage,[2] there is no doubt that the future of the nuclear navy is secure, as is the demand for highly enriched uranium.

After the 1958 agreement, Britain rapidly acquired its own facilities for making submarine reactors and their nuclear fuel. In 1959, plans were announced for a PWR fuel element factory to be built by Rolls-Royce at Raynesway in Derby.[3] By 1962, research and design offices for Rolls-Royce and Associates had been opened on an adjoining site and the complex had acquired the "Neptune" reactor, first installed by Rolls-Royce at Harwell. Although Derby is rarely to be found on maps of Britain's nuclear industry, the complex of buildings off Raynesway is the heart of the nuclear submarine fuel cycle in Britain: Rolls Royce and Associates design and supply the reactors while the fuel itself is fabricated by their Manufacturing Division (formerly the Nuclear Components Division of Rolls-Royce plc).

Rolls-Royce have described their responsibilities for the Navy's reactors as "from cradle to grave",[4] which, considering how many people could be killed by the warheads of a missile-carrying submarine, is a cliche of stupendous irony. In fact it isn't even true – submarine reactors cores and their irradiated fuel end their days in storage at BNFL Sellafield.[5] Moreover,

the highly enriched fuel – made from uranium owned by the Ministry of
Defence – begins its life also with British Nuclear Fuels: the first stages of
producing naval reactor fuel are carried out at Springfields (as detailed in the
previous chapter). In between lies the naval nuclear fuel cycle, a sequence
of processing operations linked by a flow of radioactive materials:

Springfields (BNFL):   Conversion of military uranium to hex
          ↓
Capenhurst (BNFL):   Hex enrichment to low level
          ↓
Portsmouth, Ohio (USDOE):   Hex enrichment to high level
          ↓
Aldermaston (MoD):   Receipt and buffer storage of HEU
          ↓
Derby (Rolls-Royce Ltd.):   Fabrication of HEU submarine fuel
          ↓
Barrow, Rosyth, Devonport: Re/fuelling locations of submarine reactors
          ↓
Sellafield (BNFL):   Irradiated naval reactor fuel store

A flask containing a new submarine reactor core (which until the mid-1980s were
transported by rail) passing through Derby station.

After partial enrichment by BNFL, Capenhurst's output is shipped to the Portsmouth Gas Diffusion Plant near Piketon, Ohio, where the proportion of U235 is increased to 97 per cent.[6] The product is returned to Britain by military aircraft and transported to Aldermaston by road. Reputedly Aldermaston holds buffer stocks of HEU to ensure that regular deliveries to Rolls-Royce can be quaranteed. Deliveries from Aldermaston to Derby are also made by road, and in the 1980s were taking place about every two months.

Fuelling the nuclear navy

The end product of Rolls-Royce's activities at Raynesway is a nuclear submarine reactor core. The highly enriched fuel is contained within a core "barrel" which fits inside the reactor pressure vessel. With newly built boats this operation is straightforward; the core is installed during construction at the shipyard. Once in place, however, the reactor is sealed within a massive steel containment vessel which becomes intensely radioactive. The removal of irradiated fuel is therefore time-consuming and hazardous and carried out only during lengthy refits. With the boat in dry dock, a large hole must be cut in the hull to allow the old core to be removed and a new one emplaced.

Nuclear submarines have been refuelled at three nuclear dockyards in Britain: Chatham (now closed), Devonport near Plymouth, and Rosyth on the Firth of Forth. Devonport refits the Swiftsure and Trafalgar classes, while Rosyth handles the remaining hunter-killer submarines and the Polaris fleet. Rosyth will also be responsible for refuelling the Trident submarines. In 1986, the MoD announced that work had started on a new refitting complex for both Trident and the "hunter-killers".[7] New cores are therefore transported from Derby to four locations in Britain: to Barrow-in-Furness, to be installed in submarines under construction; to Devonport and Rosyth, for submarines being refuelled; and very occasionally to HMS Vulcan.

The number of new cores leaving Derby over the years has increased as the nuclear fleet has grown. Nevertheless, as the frequency corresponds to the rate of fuelling and refuelling, it still only amounts to about one or two movements per year. The life of submarine reactor cores has gradually been extended over the years with the development of new longer life cores – from the original NR2 design through successive generations known as Core B and Core Z. By the mid-eighties, reactor cores were lasting approximately five to six years before needed refuelling became necessary.[8]

For the Trident submarines, however, a substantial increase in power was required, well beyond any stretch capability of existing reactor designs. The result was a decision in 1977 to develop the "dispersed PWR" – the PWR2 – whose prototype was installed at HMS Vulcan in 1985. The PWR2 core will have a life of seven or eight years but will not be backfittable to other submarines.[9] With a slimmed down fleet (and longer times between refits) the number of core movements in the future will decrease slightly.

A Rolls-Royce container on its way to Devonport dockyard, Plymouth. The company refuse to identify the container or its contents, but it might be assumed to be part of a nuclear submarine reactor core.

Until the mid-1980s, new reactor cores – loaded with fuel – were moved by British Rail from St. Mary's goods yard in Derby. The transport arrangements reflected the strategic sensitivity of submarine cores - not only are they radioactive, but they are fuelled with virtually the same type of uranium used in warheads. The journey from Raynesway began by road with the core transported in a cylindrical flask on the back of a lorry. A convoy of Rolls-Royce vehicles and police cars accompanied the load to St. Mary's where it was transferred by crane to a purpose-built 22 tonne railway wagon. Identified as a "Nuclear Flask Transporter" and numbered MODA 95781, it had a gross laden weight of 80 tonnes and was used solely for journeys to and from St. Mary's.

Around 1985, however, fundamental changes were made in the way reactor cores were moved. Instead of being transported ready assembled and loaded with fuel, cores are now dismantled prior to dispatch and transported in "modules" by road. Rolls-Royce refuse to divulge details of the new arrangements – they will not even confirm the mode of transport – but the company uses a variety of transport containers for different types of radioactive material and it might be assumed that the fuel is now moved separately from the actual core.

Irradiated fuel is transported from the naval dockyards to Sellafield by rail. Although movements occur infrequently – like the delivery of new fuel, they are tied to the frequency of refits – the unusual wagons involved, the conspicuous security arrangements and a 35-mph speed limit make the train stand out a mile; the transport of other types of irradiated fuel looks almost casual by comparison. The irradiated core is transported in a large flask slung in the middle of a massive railway wagon numbered MODA 95780. The "Hot Core Transporter" is one of the largest wagons on the railway network. Weighing 94 tonnes, it is also one of the heaviest. With a gross laden weight of 191 tonnes (the flask and core must therefore weigh about 100 tonnes), it is supported at each end by a pair of 3-axle bogies providing a total of 24 wheels. No less distinctive is the formation of the rest of the train. The flask carrier is flanked by two converted coaches used as observation saloons from which an escort team comprising armed MoD police and officials can keep an eye on the load in between.

The massive 24-wheeled "Hot Core Transporter" used for moving irradiated submarine reactor cores to Sellafield.

"There is no conceivable accident . . ."

Rolls-Royce and Associates's nuclear activities involve transporting a variety of radioactive materials. According to a company brochure: "RR&A has been engaged in the management of radioactive material transport for nearly 30 years. The goods carried cover a spectrum of categories including: samples and foils; manufacturing wastes; contaminated and irradiated plant components; resins and liquid effluent; sources; new and spent fuel, and decommissioning and disposal material".

Undoubtedly the transport of new and spent submarine fuel has been the most sensitive. Although there are relatively few movements per year, consignments of submarine fuel – new and irradiated – make lengthy journeys across Britain. Reactor cores from Devonport, for example, must pass through Bristol and the West Midlands. Questions about safety and the likelihood and consequences of an accident are entirely justified but totally unanswered. In 1985, for example, Derbyshire County Council's Public Protection Committee, concerned about the transport of radioactive materials from the Raynesway factory, raised their concerns with Rolls-Royce and got an all too predictable response: "The responsibility for movement rests with the Ministry of Defence and it has been a long-standing policy under successive governments not to publish such information."[10]

In fact, Rolls-Royce even refused to give the Council information which had already been published. "For reasons of National Security", the company wrote, "we do not comment on the destination or frequency of these movements" – yet the fact that submarine fuel is made at Derby and delivered to Barrow, Devonport and Rosyth was already in the published records of Parliament.[11] Rolls-Royce also refused to tell the Council where the uranium for their fuel came from – Aldermaston – even though this information had been published in their own submission to a 1980 Select Committee report.[12] Predictably, Rolls-Royce regards transport as totally safe: "There is no conceivable accident which could put the people of Derby at risk".[13]

Given that the company refuse to reveal any details whatsoever about the design of the containers used, their testing or transport arrangements, such confidence is totally insubstantiated as far as the public is concerned. However, from the limited information available, it is clear that special hazards do exist which beg legitimate questions about safety. Irradiated fuel is transported in a "Hot Core Container" fitted with a heating jacket, details of which are unavailable. No independent assessment can therefore be made of its reliability. Submarine reactor fuel is encased in zirconium which presents a fire risk under certain circumstances. Submarine reactor fuel plates are also much thinner than civil nuclear fuel elements, raising questions about how they would survive an impact if a crash occurred.

These are not academic queries – transport accidents have already occurred involving irradiated submarine fuel flasks. On 7 April 1977, when nuclear submarines were still being refitting at Chatham, the Hot Core Transporter derailed in Gillingham on its way to the nearby dockyard with an empty flask. According to the yard's nuclear power manager at the time, even an accident to a loaded flask would not threaten safety. As the manager explained to the local newspaper: "Should the inconceivable happen, such as the container being struck by particularly powerful lightning, there would still be no danger to the public. A small area of contamination around the container would have to be kept clear, but everyone a short distance away would be perfectly safe".[14]

If this implies that lightning is the most serious mishap that can befall a flask – never mind fires or crashes – then clearly there is a glaring deficency in the the International Atomic Energy Authority tests, which make no mention of such a risk nor recommend any appropriate "lightning test"! Moreover, lightning is not supposed to strike twice. Yet just six weeks later on May 26, the Hot Core Transporter was derailed again in Gillingham – and at the very same place. Although on both occasions the wagon and its flask remained upright and undamaged, railway accidents can be much more serious. What would the consequences then be, and what would be the emergency response?

The first potential problem could be identification. The 24-wheeled wagon has variously been described in the Press as carrying nuclear warheads[15] or nuclear waste[16] both of which would contain radioactive materials quite different from submarine fuel. In practice, of course, it's doubtful whether anyone would be officially told anything. In 1979, in the year before the Devonport refitting complex was opened, British Rail's Divisional Office in Bristol were informed of a contingency plan for an accident to a submarine reactor core "in transit'. In an internal memorandum to colleagues at Paddington in August of that year, the secrecy was explicit:

> Because of the obvious confidential nature of the whole exercise, it is my opinion that the B.R. Traffic Inspector should be the only person in possession of the full requirements plan A, B, C & D. This information would be handed to the nominated inspector prior to the date of the movement which he would be responsible for, in a sealed package with direction that it would be read and understood by him alone and then resealed and retained on his person until handed back to Divisional H.Q. immediately after the transit has been completed.

Precisely what plans "A, B, C and D" are has remained a secret – not only to the emergency services and local councils through whose areas the trains pass, but also to the rest of British Rail.

Research reactor fuel cycle

Research reactors have been operating in Britain since 1947. Like other reactors, they generate movements of radioactive materials; new fuel must be delivered and irradiated fuel and radioactive wastes removed. Compared to commercial nuclear power stations, however, these movements are qualitively and quantitively quite different. On one hand, research reactors are much smaller than power stations and less material requires transportation. On the other, many research reactors use fuel manufactured from highly-enriched uranium which, per unit weight, is more radioactive than most power station fuels and constitutes a greater security risk. At the time of writing (1991), 10 research reactors were operating in Britain, six of which use uranium enriched to around 90 per cent U235 (see table 5.1).

Table 5.1: British research reactors

Reactor	Owner	Location	Fuel
Nestor	UKAEA	Winfrith	20-90% U
Dimple	UKAEA	Winfrith	3 & 7% U
Viper	MoD	Aldermaston	37.5% U
Jason	MoD	Greenwich	90% U
Vulcan	MoD	Dounreay	90% U
Neptune	Rolls-Royce	Derby	90% U
UTR-300	University of Glasgow	E. Kilbride	90% U
Universities Research Reactor	Universities of Manchester and Liverpool	Risley	90% U
Consort	Imperial College	Ascot	80% U
Triga	ICI	Billingham	20% U

Source: *Nuclear Research Reactors of the World*, IAEA, Vienna, 1986

In Britain, research reactor fuel has been manufactured and reprocessed by the UKAEA at Dounreay. Known principally as the home of the fast breeder, the site was first established in the late fifties with the construction of two reactors both of which were fuelled with highly-enriched uranium – the Dounreay Materials Testing Reactor (DMTR) and the Dounreay Fast Reactor (DFR). Fuel fabrication and reprocessing facilities were built and Dounreay became what might now be termed a "nuclear park" – effectively combining on one site the operations carried out by BNFL at Springfields and Sellafield, but using highly-enriched uranium instead. Although the original two reactors have long since closed, the HEU fuel fabrication and reprocessing plants have continued to operate, providing fuel cycle services for some of Britain's research reactors and others overseas. Supplies of highly enriched uranium have been obtained either from the United States or – mostly – by recycling material recovered from reprocessed research reactor fuel. These activities have generated movements of highly enriched uranium in various forms, principally:

1 Fuel elements or fuel "plates", manufactured at Dounreay from HEU metal, dispatched to research reactors in Britain and overseas.
2 Irradiated HEU fuel, returned from research reactors in Britain and overseas to Dounreay for reprocessing.
3 Highly-enriched uranium metal, for fabricating new fuel elements, has been delivered very occasionally to Dounreay from the United States. (Only one order was received during the 1980s).

Domestic movements of highly-enriched uranium research reactor fuel

The transport of research reactor fuel varies enormously according to the size of the particular reactor. Used only for experimental purposes, a research reactor's fuel inventory need only be big enough to sustain a chain reaction. Movements of new and used fuel are few. The Navy's Jason reactor in London burns up less than one gram of uranium per year and according to the Ministry of Defence, is still operating ". . . on substantially the same fuel with which it started life . . .", back in 1962.[17] Although some surplus fuel was returned to Dounreay for reprocessing during the early 1970s, the amount moved in and out since has been minimal.

Other research reactors, such as those used to manufacture radioactive isotopes, are much larger. For many years, the largest in Britain were the DIDO and PLUTO Materials Testing Reactors (MTRs) at Harwell and the Herald reactor at Aldermaston. Until their closure in 1990 and 1988 respectively, deliveries of new and used fuel between these reactors and Dounreay accounted for most movements of research reactor fuel in Britain.

Compared to the movement of fuel for nuclear power stations, the transport of research reactor fuel has largely escaped public attention. Movements are relatively infrequent and the use of highly enriched uranium ensures a high level of secrecy. Occasionally, however, details emerge when something goes publicly wrong. In 1981, for example, an emergency landing at Aberdeen Airport forced the UKAEA to reveal a few details about the transport of new MTR fuel to Harwell. These were being delivered about once a month by air. At 11 a.m. on October 19, a twin-engined freight aircraft belonging to Air Ecosse put down at Aberdeen with a suspected undercarriage fault. Although the fault turned out to be nothing worse than an instrument failure, the police and the fire brigade cordoned off the plane while several 8' x 5" containers of fuel were offloaded. After being transferred to another plane, the fuel, bound for Harwell, continued its journey to RAF Abingdon in Oxfordshire – conveniently near to the UKAEA site.[18]

Deliveries of irradiated fuel to Dounreay are made using the "Unifetch" flask, specially designed by the Authority for transporting MTR fuel. A cylindrical-shaped flask incorporating 300 mm-thick steel sides, it weighs just under 17 tonnes unladen. It has a capacity of between 26 and 40 fuel elements, depending on the design of the fuel.

A "Unifetch" flask, transporting research reactor fuel from either Aldermaston or Harwell, approaches the M74 on its way north to Dounreay.

Since the closure of the Harwell and Aldermaston reactors, use of the Unifetch has declined. When the Harwell MTRs were operating, for example, irradiated fuel was transported to Dounreay about nine times per year on average. With deliveries confined to the summer months, the lengthy journeys north via the M6 and M74 were broken up by overnight stops at BNFL sites close to the route. Both Springfields and Chapelcross have been used for this purpose.

By 1991, however, Harwell and Aldermaston had dispatched their last deliveries of MTR fuel. In Britain, the Unifetch is now used mainly for occasional movements of fuel from the University reactors at Ascot, Risley and East Kilbride. It is also used for imports of fuel to Dounreay and has been hired by customers in the United States for transporting fuel within North America.

Imports and exports of highly enriched uranium

Movements of highly enriched uranium are not confined to domestic routes. Military demand and the fuel requirements of research reactors generate international shipments of the material in various forms. In Britain, these have consisted mostly of imports of HEU metal from the United States, principally for use as submarine fuel.

These deliveries are the result of the 1958 US-UK Mutual Defence Agreement which enabled the two countries to trade nuclear materials for military use. While the quantities involved are secret, it has been estimated that since 1958, some 3 or 4 tonnes of HEU have been imported by Britain for military purposes. Most will have arrived during the 1960s for use in nuclear warheads, although at least 1.5 tonnes have probably been imported since for submarine fuel.[19] In 1981, for example, the MoD signed a five-year contract for HEU from the US DoE. Reportedly the eighth such contract since high enrichment ended at Capenhurst, it was sufficient to fuel 16 nuclear submarines and one land-based prototype.[20] One calculation has estimated that the amount of enrichment contracted for (100,000 swu) would provide 338.5 kgs of highly enriched (weapons-grade) uranium per year.[21]

The 1981 contract would have expired in 1986, and supply arrangements since then have been unclear. A new contract for HEU was signed in 1987, although it covered only a single shipment of metal from the US government stockpile. The material, from the Y-12 weapons plant at Oak Ridge, Tennessee, was to be delivered during 1988 for military purposes. The one-off nature of this purchase suggests that it might have been for use in nuclear weapons.

Highly enriched uranium has also been imported for use as research reactor fuel although few deliveries have been made since the 1970s. US NRC export licences were issued between 1971 and 1973 for the export to Britain of some 327.71 kgs of 93 per cent HEU for use in the UKAEA's research reactors. Since then, however, licences for only two substantial quantities have been issued for deliveries to Britain: in 1974, for 19.4 kgs of 93 per cent HEU for the (now closed) Dragon reactor at Winfrith; and in 1984, for the export of 27.5 kgs of 60 per cent HEU for fuel fabrication by the UKAEA for the Hifar reactor in Australia.

Virtually no information is officially revealed regarding transport arrangements for imports of highly enriched uranium. However, US deliveries to other European countries give a few insights into how HEU crosses the Atlantic. In 1986, for example, over 300 kgs of HEU were flown to Europe for use in experimental French and West German reactors. The highly enriched material was transported by a US DoE "Safe Secure Transport" (SST) vehicle (see also Chapter 17) to a US Air Force base for collection by French or West German airforce planes.[22] Similar arrangements were employed in 1988, when just over 100 kgs of HEU were supplied to Canada for use in research reactors. The highly enriched material was flown to Canada from Albuquerque air base, New Mexico (also the communications headquarters of the SST system) by a Canadian military aircraft.[23] The Albuquerque base was also used in 1988 when the French air force collected 180 kgs of HEU for five European research reactors.[24]

Most of this traffic, including US supplies to Britain, consists of HEU in the form of uranium metal. However, HEU has also been imported and exported by Britain in the form of fabricated research reactor fuel. This traffic results from UKAEA contracts to manufacture and reprocess fuel for overseas research reactors: new fuel is exported from Dounreay, irradiated material is occasionally returned for reprocessing. During the sixties and seventies, HEU fuel was exported to reactors in Denmark (Riso), Australia (Lucas Heights), West Germany (Julich), India (Aspara), and South Africa (Safari-1). Fuel elements were first delivered to Harwell from where they were dispatched by air from Heathrow to Europe, or by sea to other continents.

Exports of new fuel were duly followed by imports of irradiated fuel for reprocessing. The first came during September 1962 from the DR3 reactor at Risø in Denmark. Twenty five irradiated fuel elements of 80 per cent enriched uranium arrived at the Scottish port of Leith on the *m.v. Finland* from Copenhagen. In the following year, Dounreay received its first shipment of HEU fuel from Australia: 150 fuel elements from the Hifar reactor in Sydney arrived at Liverpool's North West Gladstone dock from Melbourne. After these early deliveries, imports of irradiated HEU fuel expanded. Dounreay's reprocessing contracts included, for example, fuel from Aspara and Julich and from the Pégase reactor at Cadarache in France.

Until recently, however, such trade had declined; many of the UKAEA's reprocessing contracts expired and there are fewer reactors worldwide using HEU fuel. Some research reactors (for example, at Riso and Julich) sent their irradiated fuel to the United States instead. After 1972 no overseas MTR fuel was reprocessed at Dounreay.

There has also been pressure from the United States government to discourage the use and transport of a fissile material from which nuclear warheads can be made. The possibility that highly enriched uranium might be stolen or "diverted" by national organisations or terrorists for use in

weapons led the US Department of Energy to set up the Reduced
Enrichment Research and Test Reactor programme (RERTR) to develop
low-enriched fuels (preferably no more than 20 per cent enrichment) which
could be used as a substitute by research reactors. Outside the United States,
a number of operators have converted their reactors to low enriched fuels.
The Australian Atomic Energy Commission's Hifar reactor near Sydney
provides a recent example. After its last purchase of HEU (supplied by the
US but manufactured into fuel at Dounreay), Hifar was converted to use low
enriched uranium.

With the beach in the background, two Goslar flasks of irradiated fuel from the
PTB research reactor at Braunschweig, Germany arrive in Dover off the overnight
Dunkerque train ferry.

At the end of the 1980s, however, Dounreay's overseas business began
to revive. With the closure of research reactors at Harwell and Aldermaston,
Dounreay looked overseas for new contracts to replace lost business.
Prospects were boosted by the imposition of an embargo on importing
research reactor fuel into the States. In December 1988, opposition groups in
the States forced the US Department of Energy to declare a moratorium on
such imports pending an environmental assessment of the transport hazards.
The problems this caused for various research reactor operators around the
world – who would otherwise have sent fuel to the States – created a timely
business opportunity for the Dounreay reprocessing plant.

With almost 80 research reactors in 33 countries using highly enriched uranium fuel supplied by the United States or other western countries[25] business looks promising. In 1989, the UKAEA signed a contract with the German company KFK to reprocess uranium/thorium residues. This was followed by the announcement of a deal to reprocess German fast reactor fuel pins which had been irradiated in a French reactor. Further contracts were being negotiated to reprocess fuel from research reactors at Petten in the Netherlands, near Madrid (the JEN reactor) and in Berlin (at the Hahn-Meitner Institute). In 1991, it also emerged that Dounreay expected to receive highly enriched uranium from the Iraqi research reactors bombed during the Gulf War. Apart from increasing business for the UKAEA, such contracts will also increase public concern.

Issues

Dounreay's new contracts to reprocess research reactor fuel show every sign of generating the sort of controversy that has plagued BNFL's operations at Sellafield. Not only do they bring more imports of irradiated fuel to Britain but also the familiar concerns about health and safety and wider issues about reprocessing in general.

One of the new contracts, to receive fuel from a research reactor at the Hahn-Meitner Institute in Berlin, has already provoked opposition in both Scotland and Germany. During 1990, in what was formerly West Berlin, the ruling "Red-Green" coalition between the majority socialist members of the senate and their partners from the Alternative Liste (the local green party) broke up after the Environment Senator, Michaele Schreyer, of the Alternative Liste, refused to sign a license authorising the HMI reactor to start up after being closed for refurbishment. Some members of the senate had expressed concern that the proposed method of waste disposal – sending it to Scotland for temporary storage and possible reprocessing – was unsafe.

Similar concerns were expressed in Scotland. A delegation representing Scottish Nuclear Free Zone local authorities travelled to Berlin to state their opposition to the deal. The delegation's leader, Lothian Regional Councillor Will Herald, told Senatorin Schreyer: "There are fears in Scotland that such a development would open the doors for other nations to send their waste to Dounreay for storage and reprocessing, effectively turning the North of Scotland into the nuclear dump of the world". Councillor Herald also bemoaned the lack of democracy, pointing out that: "No consultation was carried out with the electorate of Scotland as to whether or not they wanted to become a nuclear waste dumping ground for Germany, Europe or the rest of the world". Suspecting that the Berlin contract could be "the tip of a very large iceberg", the delegation emphasised its view that irradiated fuel generally should not be transported offsite for reprocessing but stored instead at the point of production.

Another chunk of the iceberg surfaced in 1991. After the AEA had signed a contract with another German research institute, the Physikalisch Technische Bundesanstalt (PTB) at Braunschweig (for interim storage of irradiated research reactor fuel with a reprocessing option) two flasks of highly enriched uranium fuel elements were duly dispatched to Britain. The flasks got as far as Rotterdam docks, where union members affiliated to the International Transport Workers Federation refused to load them onto a regular shipping service to Liverpool. The workers' action followed a union decision not handle nuclear materials unless they were transported on specially designed ships. In Britain, the Braunschweig delivery was criticised by national and local politicians. The leader of Lancashire County Council, through whose area the fuel would have travelled if it had arrived at Liverpool, condemned the shipment while the Scottish Labour Party demanded a halt to all imports of "nuclear waste". After being stranded in Rotterdam for five days, the Dutch Environment Minister eventually ordered the flasks back to Germany. The consignment eventually made it to Scotland in October 1991, via the Dover-Dunkerque train ferry. After travelling first to Winfrith by rail, the rest of the journey was completed by road. Protests against the delivery included a vehicle blockade near Inverness which briefly brought the flasks to a halt.

Such movements inevitably raise concerns about health and safety. Like the transport of irradiated fuel from power stations, the risks and consequences of an accident were foremost in critics minds. According to Scottish National Party vice-convenor Mike Russell: "The transportation of highly radioactive materials over long distances by rail, road and sea puts at risk all those communities through which they pass. Even the threat of a nuclear accident will do immense damage to industries which rely on a clean environment such as farming, fishing and tourism."[26] If road deliveries of irradiated fuel from British research reactors had been confined to the summer months of the year – to avoid the transport risks associated with Scottish winters – why change practice for imports?

Apart from worries about accidents, there have been problems even during routine movements of such fuel. On 1 February 1991, the AEA received a Unifetch flask containing irradiated fuel from a research reactor at the Bhabha Atomic Research Centre in India. The fuel had originally been lent to India by the UKAEA after a 1964 contract. When the flask was monitored at Dounreay – for the first time since sailing from India six weeks earlier – radioactivity was found on its outside surface and also on the flat-bed transporter which had brought the flask by road from Felixstowe docks. Moreover when, as a precaution, the Authority checked the container base on which the flask had arrived at Felistowe – and which by then had been sitting in the port for two weeks – that too was found to be contaminated and had to be removed to Harwell for decontamination.[27] Even if there was no risk, such incidents suggest that the movement of research reactor fuel is set to become a growth industry – for the critics as well as the UKAEA.

Hazards

Highly enriched uranium is considerably more toxic and radioactive than natural or low-enriched uranium; high enrichment increases the proportion of the more hazardous isotopes of uranium, U234 and U235. The isotopic composition of natural and 90 per cent enriched uranium is compared in Table 5.2.

Table 5.2: Isotopic composition of natural and highly enriched uranium[28]

Isotope	Content (weight per cent)	Specific α-activity decays/μg mixture	Total α-activity (per cent)
Natural U			
U234	0.054	0.737	49.07
U235	0.7115	0.028	1.86
U236	–	–	–
U238	99.283	0.737	49.07
90 per cent enriched U			
U234	0.87	119.192	96.3
U235	90.00	4.285	3.5
U236	0.17	0.244	0.2
U238	8.96	0.067	0.0

As U235 is 100 times more alpha active than U238,[29] the toxicity of highly enriched uranium – which may consist of up to 97 per cent U235 – is correspondingly greater. The enrichment process also increases the proportion of U234, which is even more alpha active than U235. The increase in this isotope, which occurs only in minute amounts in natural and low enriched uranium, makes highly enriched uranium sufficiently radioactive to require handling only in glove boxes.

Solid uranium metal oxidises in air at room temperature. Over a period of three or four days, a film of oxide forms on the surface. If finely divided, uranium metal is pyrophoric. At a temperature of 700°C or higher, it will burn in air or oxygen emitting a blinding white light. The result is a black powder of U3O8.

After irradiation, highly enriched uranium fuel elements are considerably more radioactive (per unit weight) than those made from natural or low enriched material. With a different isotopic composition, the fission of highly enriched uranium fuel produces a different range of by-products to those created inside natural or low enriched uranium fuel. The high percentage of U235 allows relatively high burn ups to be achieved – in some cases up to 50 per cent – thus creating a higher proportion of fission products than present in other uranium fuels. Conversely the smaller amount of U238 present means that far less plutonium is created by irradiation. For example, irradiated fuel elements from the Hifar research reactor in Australia (60 per cent enriched uranium) contain about half a gram of plutonium and 89 grams of U235.[30]

6   Spent fuel – domestic

Two days before the forty fourth anniversary of Hiroshima, Bob Cole and Rod Stallard went train spotting. Both were active in the Welsh CND (CND Cymru) campaign against the proposal to build a PWR at Wylfa. The plan to build a second nuclear power station on Anglesey had been widely criticised: 20,000 people registered written objections with the Department of Energy.

Cole and Stallard's response was more forthright. Once a week, flasks of spent nuclear fuel from the existing magnox reactors at Wylfa travel along the North Wales coast on their way to Sellafield. The journey is broken at Llandudno Junction where flasks from Wylfa join those from neighbouring Trawsfynydd. On Friday, 4 August 1989, the two men walked across the tracks at Llandudno Junction, climbed on to a train from Anglesey and chained themselves to a wagon transporting a flask.

This was not the first demonstration against flasks of irradiated fuel which travel from Britain's nuclear power stations to Sellafield. Several years before, demonstrators in Leiston, Suffolk held up a train transporting a flask from Sizewell (on a line used by no other trains) by the simple but effective action of sitting on the road over a level crossing and refusing to budge – if the gates can't be moved, neither can the train.[1] Other demonstrations have occurred at London stations through which flasks of spent fuel are transported.

Health and safety issues have always been a major concern. What is popularly known as "nuclear waste" and technically as "irradiated nuclear fuel" – consists of the used fuel elements that are removed from a nuclear reactor. During their sojourn in a reactor, fuel elements undergo nuclear "fission", a process which not only creates heat to produce electricity but also a range of highly toxic and radioactive by-products which accumulate inside the fuel. These comprise new substances, such as plutonium, and "fission products", the radioactive elements formed by splitting atoms of uranium. The most hazardous include krypton-85, strontium-90, ruthenium-106 and caesium-137. Only a tiny percentage of spent fuel – the unwanted fission products – can properly be called waste. The rest consists of uranium

– up to 99 per cent – which can be recycled, and plutonium. A full flask of AGR fuel will typically contain about 840 kgs of uranium and 3.7 kgs of plutonium.[2] Magnox fuel flasks usually contain about 4 kgs of plutonium.[3] Uranium and plutonium are recovered by reprocessing at Sellafield. With a radioactive inventory of up to five *million* curies (37,000 TBq) per flask, the possibility of an accident or sabotage causing a radioactive leak has been a persistent fear.[4]

A magnox flask from Hinkley Point A passing through Bridgewater, Somerset on its way to the train to Sellafield.

Even under normal transport circumstances, loaded flasks emit radiation. Although the amount is relatively small, gamma radiation penetrates even the thickest flasks in use and irradiates the surrounding environment. Flasks also have a history of contamination on their outside surfaces, a problem which can result in small quantities of radioactivity being released into the environment. (These issues are covered in more detail in Chapter 13.)

The Welsh action, however, showed that wider issues are involved: transport hazards are not the only reason why the movement of spent fuel is contentious. The timing of the Llandudno Junction protest – close to the anniversary of the Japanese bombings – was no coincidence. Reflecting concern that plutonium produced by nuclear power stations might be used in nuclear weapons, the theme of Welsh CND's campaign was "Wylfa B is for Bombs". Slogans to that effect were daubed on the flask wagon at Llandudno Junction while the campaign leaflet juxtaposed photos of flasks

with Trident missiles and mushroom clouds. Inside, Welsh CND quoted
Michael Barnes QC, the government inspector at the Hinkley Point Inquiry
that had ended earlier that year: "there can be no safeguard in the long
term that such plutonium will not be used for military purposes, except not
constructing the reactor . . ."

The possibility that plutonium from Wylfa B might be used for non-
peaceful purposes was one of the reasons why Gwynedd County Council
– whose area includes Wylfa and Llandudno Junction – opposed the
application for a new reactor. The military overtones of civil nuclear
power were also acknowledged by the Fire Brigade who were called
out by British Rail to cut the protesters from the train: when Cole and
Stallard pointed out that they, like the Fire Brigade, were members of
trade unions affiliated to CND, the Fire Brigade officers declined to sever
their chains.

The Welsh have other reasons for being sensitive to the nuclear industry.
In 1990, some 300,000 sheep, 416 holdings and 210,000 acres of Welsh
land were still subject to restrictions imposed after being contaminated with
radioactivity from the Chernobyl accident four years earlier.[5] Elsewhere,
Wales is invaded by radioactivity from Sellafield as a result of reprocessing
spent fuel, including the contents of those flasks which travel along the
North Wales coast. The extensive flooding on that coast in 1990 showed
that radioactivity was returning to Wales in sediment washed ashore; when
it dried out to be blown by the wind, local people might well have inhaled
radioactive particles that had previously travelled along the railway line
from Wylfa. Thus the transport of spent fuel is inextricably part of a larger
problem. There is also a matter of principle: as a result of declarations by
its constituent county authorities, the whole of Wales is supposed to be a
nuclear free zone.

However, the major concern about the transport of spent fuel has been the
possibility of a leak: if any of the radioactive contents escaped, people could
be exposed to potentially fatal radiation. However unlikely that may be, the
question of how to cope with such an event adds to the controversy. Local
authorities and emergency services have been drawn into a debate by the
unwillingness of British Rail and the nuclear industry to provide sufficient
information for contingency planning. Having acquired a relatively high
profile, spent fuel transport has also become a symbolic campaigning target
for those opposed to other aspects of the industry.

Transport flasks – ". . . small-scale nuclear installations . . ."

Spent fuel flask designs mostly fall into two generic types: shaped like cubes or like cylinders. Most of those used in Britain are cube-shaped and can be described in the simplest terms as an open-topped thick metal box with a removable lid for loading and unloading. Inside, fuel elements are contained within a skip and the flask is filled with water to help remove heat from the radioactive fuel. Cylindrical designs are tube-shaped, with one end sealed and a removable lid at the other.

British nuclear power stations currently make use of four basic types of flask. (Irradiated fuel from Sizewell B and any other future PWRs will use Excellox flasks, described in Chapter 7.) Each type is generally used for a different type of fuel, as follows:

Magnox

Used by all Nuclear Electric and Scottish Nuclear Magnox power stations except Chapelcross and Wylfa (see below).

An AGR A1 flask from Hinkley Point B passes through Bridgewater on its way to the train to Sellafield.

AGR (A1)

A variation on the standard magnox cuboid design. The A1 is the original AGR flask, externally distinguished from its magnox counterpart by a prominent shock absorber bolted onto the lid. AGR fuel is ten times more radioactive per tonne than magnox fuel, so the A1 flask has much thinner steel walls – 3.5 inches – but incorporates a 7 inch internal lead lining. A1 flasks are also used for transporting fuel from the Magnox reactors at Wylfa.

Mark 2 AGR (A2)

In January 1991, the first of a fleet of 46 second-generation AGR flasks was introduced. First used by Scottish Nuclear for deliveries to Sellafield from Hunterston B, the A2 will also be used by Nuclear Electric. Unlike the original AGR flask (which will remain in service for a few more years), the A2 design is an all-steel flask with 360 mm (14.25 inches) thick side walls.

An AGR A2 flask from Hunterston B arrives at Fairlie station, Scotland, where flasks are loaded for dispatch to Sellafield.

1120

This design is a cylindrical flask, used only for transporting spent fuel to Sellafield from the UKAEA's Steam Generating Heavy Water Reactor (SGHWR) at Winfrith. As this reactor closed in 1990, deliveries from Winfrith will eventually end.

After its removal from a reactor, spent fuel continues to generate heat from the decay of the fission products. To remove this heat during transport, loaded flasks are filled with water. Heat is removed by convection to the outside surface of the flask, where it dissipates with the help of metal cooling fins welded on the outside of the flask.

Flasks are transported on special railway wagons, Nuclear Flask Carriers, otherwise known as "flatrols". In practice, the flasks themselves are rarely seen because many of the flatrols have been fitted with aluminium covers which hide them from public view. Indeed, but for small radiation symbols on the side, there is little to indicate the presence of what a government Royal Commission once called "in effect a small-scale nuclear installation".[6] This description of a flask is an apposite reminder that beneath the anodyne cover, the cargo of a flatrol wagon was once part of the very core of a nuclear reactor.

Flask safety

With the exception of a few minor derailments, nuclear flasks have suffered neither a major rail accident nor a major radioactive leak. Yet despite this record, many people are not reassured. Accidents are not the only potential hazard. The transport of spent fuel under "normal" circumstances – without an accident – is responsible for a small dose of gamma radiation to the surrounding environment. For this reason, instructions in BR's "Working Manual for Rail Staff" include the following:

E2/23   "Radioactive substances which are detained for any reason must be placed away from vehicles containing other dangerous goods, livestock, sensitised photographic materials, and from staff rest rooms, shunters' cabins and other similar places. If the vehicles are likely to be detained for 12 hours or more, the distance separating the vehicles and places mentioned must be at least 20 ft (6 metres)"

E2/25   "No (irradiated fuel) flask must be within 20 ft (6 metres) of a locomotive or a vehicle in which the guard is riding. If the distance is less than 20 ft a barrier wagon with an actual weight of more than 8 tons (8,150 kg) must be provided."

BR's conclusion about the routine hazards of transporting spent fuel is ambiguous: "flasks . . . will present no hazard to staff in the normal course of their duties, but no person must remain near the wagon unnecessarily."[7] Moreover, even in the absence of major accidents, there have been incidents of contamination with small quantities of radioactivity released to the environment from the outside surfaces of flasks (see Chapter 13).

However, most public concern has centred on the possibility of a major accident breaching a flask. The fact that such an accident has so far not happened is no guarantee that one could never occur in the future. Questions about safety are inevitable. Could a flask survive an impact or a fire? Could the contents of a flask ever leak and what would be the consequences if they did? The answers depend on the testing of flask designs.

Flask designs are tested to IAEA standards – notably the 9-metre drop test and 800°C fire for 30 minutes – which should ensure that they remain safe and intact in accidents. This might seem adequate, yet doubts have remained, especially about testing procedures which have been repeatedly criticised. Although the CEGB have used full-sized flasks in fire tests, for many years drop-testing used only models. In 1975, for example, a government Royal Commission commented: "We were surprised to learn that the tests are conducted only on models and since the containers travel on ordinary freight trains at speeds of up to twice that assumed in the tests, we were not wholly reassured."[8]

As a result of such criticism, the CEGB eventually subjected a full-sized flask to a 9-metre drop test at their Structural Test Centre at Cheddar in 1984. The demonstration was part of a £4 million four-year programme which culminated in spectacular fashion on 17 July 1984 when the same flask (but with a new lid) was hit by a 100-mph driverless train at British Rail's Old Dalby test track in Leicestershire. This public relations extravaganza was witnessed by spectators from local authorities, trades unions and the Press. Many people were clearly impressed; the water-filled flask, although slightly damaged, did not leak and the spectacular destruction of locomotive 46009 was repeated on endless "action replays". The event did not pass without criticism, however.

First, the IAEA tests stipulate that flasks should be tested cumulatively for impact and fire. The same flask used in an impact test should be used again in a fire test to simulate one of the worst kinds of railway accident, a crash followed by a conflagration. However, any possibility of further tests on the Old Dalby flask was scuppered when it was subsequently sent to the Railway Museum at York. In fact, fire tests had already been performed on flasks and, as the CEGB freely admitted, the Old Dalby crash was principally a PR exercise and not an approval test to IAEA standards. Nevertheless, it seems surprising that the CEGB didn't take the opportunity to impress the sceptics further by a literal compliance with the IAEA regulations.

Second, the crash failed to test the effects of impact on the fuel itself. The crash test flask contained only an equivalent weight of solid non-radioactive steel bars, "simulating a two-tonne load of fuel". Whilst the steel bars may have simulated the weight, they could not simulate the effects of impact on two tonnes of the real thing. This is no mere technical quibble. Any breakage of the fuel elements could seriously increase contamination if the flask contents were subsequently to leak. Spent fuel elements are considerably less robust than the flasks that transport them. The effects of intense radiation inside a reactor can also make them brittle. Magnox fuel rods are encased in a cladding whose thickness varies between 1.7 mm and 2.8 mm (different reactors have different specifications). With AGR fuel, the stainless steel fuel pins which contain pellets of enriched uranium oxide are just 0.4 mm thick. Considering that the cooling fins on the Old Dalby flask were dented and buckled by the impact – and they are about 25 mm thick – it might be expected that such an accident would damage the fuel itself. Any breach of spent fuel cladding will release radioactive gases inside the fuel flask; if an accident was severe enough to rupture the flask, gases such as krypton-85 and caesium-137 could escape to the environment.

The result of the CEGB's Old Dalby train crash; the buckled fins raised the question – how would thinner flasks have survived?

A third criticism of the Old Dalby test concerned its use of a magnox flask – many deliveries of spent fuel are made using flasks of a different design. In particular, most AGR fuel is currently transported in the A1 flask which is significantly different: while magnox flasks are 370 mm thick, the original AGR flasks have steel sides only 89 mm thick with a 178 mm thick lead lining on the inside of the walls, a feature not incorporated in magnox flasks. The function of the lining is to provide additional radiation shielding and it contributes little to structural strength: the integrity of the A1 flask depends largely upon steel sides less than a quarter of the thickness of those in the magnox design. Considering that the Old Dalby crash was staged at a time when AGR power stations were being commissioned at Dungeness, Heysham and Hartlepool – testing a magnox flask at Old Dalby did not seem entirely appropriate.

Finally, despite the CEGB's efforts at Old Dalby, even the speed of the 100-mph train crash has been criticised as inadequate. While it may suffice for a high speed impact with a stationary object, the combined speed of two trains colliding from opposite directions could be far higher. Furthermore, overall train speeds are rising with the introduction of faster new electric locomotives. (BR's Class 91, for example, has a designed maximum of 149 mph.) As rail speeds progressively increase, so will criticism of impact testing – even at 100 mph.

Such arguments must exasperate the industry. After all, it will probably always be possible to conceive of ever worsening accident scenarios, however unlikely they may be. One way out of the argument was proposed by the Ecology Party in 1980: "Instead of testing a flask merely to ensure that it does not leak up to a certain impact velocity . . . test one until it does leak, to determine the point at which its physical integrity breaks down".[9] "Testing to destruction" would determine the breaking point of a particular design.

A more fundamental criticism of the Old Dalby test was its failure to test the effects of fire. Britain's petrochemical industry produces and consumes a wide range of inflammable commodities, such as kerosene, fuel oil and liquified natural gas, much of which moves by rail. Block trains carrying over 2,000 tonnes of inflammable liquid run on lines also used by nuclear flasks. Past experience showing that hydrocarbon fires can burn for longer periods and at higher temperatures has led to criticism that the IAEA's fire test criteria are inadequate.

This was dramatically illustrated just a few months after the CEGB's crash. On 20 December 1984, a freight train caught fire inside the Summit Tunnel near Rochdale after an axle failure derailed two 100-tonne petrol tankers. The train, from Haverton Hill, Teesside, was carrying thousands of gallons of petrol to ICI's plant at Glazebrook near Warrington. Three quarters of a mile into the tunnel, automatic brakes stopped the train and the line under the Pennines became an inferno. Steel rails broke in the heat and almost a quarter of a mile of track was destroyed. Near the heart of the fire, the brick tunnel walls melted and vitrified. The blaze took four days to extinquish and according to a report in BR's newspaper, *Rail News*, temperatures reached a staggering 8,000 degrees centigrade – ten times the temperature of the IAEA fire test.[10]

Fires of such ferocity are fortunately rare, and the subsequent inquiry suggested that the maximum temperature had been lower than the figure in *Rail News* – but still well in excess of the IAEA's test criterion. But even if those statistics were unique, accidents that end in flames are not. In the previous year, for example, a tanker train from the Stanlow refinery in Cheshire derailed and caught fire near Acton Grange Junction near Warrington – the junction where spent fuel travelling to Sellafield from North Wales joins the flask route from the south. The nearby sidings at Walton Old Junction are sometimes used as a "layover" where flask trains break their journey to the north. An eyewitness to the derailment recalled: "There were three tremendous explosions. About a quarter of a mile of the embankment was ablaze after oil poured down it."[11] The location of the fire had another hazardous significance – it was just 600 yards from a Dupont chemical distribution centre.

Such accidents beg an obvious question: what would be the effect of a serious fire on a nuclear flask? The physical effects should be beyond dispute; determined as they are by the materials involved. Prolonged exposure to high temperatures would eventually boil the cooling water inside a flask, and if high temperatures were maintained for long enough, the fuel itself would melt. Uranium metal (magnox fuel) melts at 1,130°C; magnox fuel cladding at 645°C; and uranium oxide (AGR fuel) at 2,800°C. However, the *consequences* of these effects are more contentious. Critics have suggested that the resulting increase in pressure could potentially rupture a flask or force a leak through the seals or bolts, the relatively weak points of flask construction (especially if these components had themselves been strained or damaged by a crash). If the fuel elements were damaged or had melted, highly radioactive material could be released into the flask cavity, and, if flask integrity were breached, escape to the outside environment. A combination of impact and fire is clearly the worst of all scenarios: cooling water leaking through a breach would be replaced by air, thus allowing the fuel itself to catch fire if exposed to a high enough temperature. In these circumstances, and depending on the size of the breach and prevailing weather conditions, a full scale radioactive leak might occur.

The industry view, on the other hand, is reassuring. A CEGB information leaflet on the transport of spent fuel dismissed fire as a hazard: "With 14-inch steel walls it would take several hours for the heat to penetrate to the used fuel rods. A fire would not therefore increase the risk of radioactive release."[12] The fact that other flasks used in Britain have steel walls much thinner than 14 inches is conveniently ignored – clearly, it would take less time for heat to penetrate the 3.5 inches of steel in an AGR A1 flask.

One outcome of the CEGB train crash was a leak of a different sort. Confidential documents obtained from the Board's engineering consultants, Ove Arup and Partners, revealed that an impact from a high speed train – as demonstrated at Old Dalby – had *not* been regarded as the most damaging accident that a flask might experience. Arup's analysis showed that other types of accident were regarded as greater potential risks. These included flasks colliding with bridge or tunnel abutments or falling to the ground from high bridges. The common factor in these situations is a collision with an immovable object which is more resistant than a flask. By comparison, a colliding high speed train falls apart on impact, thus dissipating its kinetic energy – as the Old Dalby test showed.

According to Arup, the worst kinds of accident were even "likely to cause complete detachment of the lid from the body". This startling observation was not a hypothetical prediction. Earlier work carried out by the consultants had included dropping model flasks onto the granite floor of Merrivale Quarry, near Tavistock in Devon. The results were described in a confidential 1982 report by Ove Arup and the CEGB. When a quarter-scale Magnox flask was dropped from a height of 36 metres (Test No. 2), all 16 lid bolts broke and the lid separated from the body. Three years later, in a Channel Four documentary, the CEGB's Brian Mummery, in charge of transporting spent fuel, was asked "Have all the flasks you have tested come through without failure?". Mummery's reply was short and unambiguous: "Yes".[13]

The failure of the flask at Merrivale Quarry helps to explain part of the consultant's later work for the CEGB, which involved identifying all railway bridges higher than 9 metres on the routes used by spent fuel flasks - a total of 55. Topping the list with a height of 50 metres is the Ballochmyle Viaduct near Catrine on the Scottish line from Ayr to Gretna Junction. This forms part of the flask route to Sellafield to the Hunterston nuclear power stations. Several other bridges on the same stretch of line, and on the route to Torness, are also several times the height of the IAEA's 9-metre drop test criteria (as used by the CEGB in their public test at Cheddar). Furthermore, the geomorphology of Scotland not only affords the highest potential drops for a falling flask, but also some of the hardest rocks underneath. If the bolts of a *model* flask will shear when dropped from a height of 36 metres, how would a *real* flask behave? In reality, the ground below many high bridges would not be solid. The Ballochmyle viaduct, for example, spans the tree-lined gorge of the River Ayr; other viaducts cross fields and softer rocks like sandstone. Nevertheless, Arup's analysis suggested that whatever the electricity companies may say in public, there are still residual risks which the industry are prepared to live with – the transport of spent fuel will only cease to be controversial when other people are prepared to live with them as well.

Transport routes

With the exception of flasks from Chapelcross which travel to Sellafield entirely by road, spent fuel from British nuclear power stations is transported by rail. Weekly timetabled services are run and almost always by "dedicated" trains. These are freight services run solely for the electricity generating companies that carry no other cargo. At three power stations – Winfrith, Heysham and Hartlepool – sidings run onto the reactor site and enable flasks to be dispatched directly by rail. Elsewhere, flasks must be transported by road to the nearest railway siding with a suitable crane.

At several places along the routes to Sellafield, flasks travelling from different power stations are joined up to form a single train. Flasks from Sizewell and Bradwell join Dungeness traffic at Willesden Brent sidings. Flasks from Berkeley, Oldbury and Hinkley Point are marshalled together at Gloucester, whilst those from the Welsh nuclear power stations connect at Llandudno Junction.

In addition to regular movements to Sellafield, sample quantities of irradiated fuel from power stations are periodically transported by rail to Winfrith or the Berkeley Nuclear Laboratories for "post-irradiation" analysis. Such cross-country movements are irregular (519 flask movements between 1962 and 1989[14]) but take flasks through places like Southampton which they would not normally visit.

With Southampton Tunnel and the town hall clock tower in the background, two spent fuel flasks (hidden by aluminium covers on the flatrol railway wagons) approach Southampton station on their way to Winfrith.

Transport patterns could change substantially in the future if the generating companies reconsider and "backend" arrangements. One possibility – a proposed dry "buffer" store for AGR fuel – was announced in 1986. The joint project between the CEGB and the SSEB was to be sited at Heysham nuclear power station.[15] A buffer store would provide the utilities with more flexibility in spent fuel management especially if there were hiccups in operations at Sellafield. The store would change transport routes for AGR spent fuel: flasks from AGR power stations would deliver fuel to Heysham rather than Sellafield. After interim storage, the fuel would be transferred to Sellafield.

However, by 1990, enthusiasm for the joint store had waned with both companies considering the possibility of separate stores. In 1991, Scottish Nuclear announced plans for "modular spent fuel dry storage facilities" at Torness and Hunterston "B", a proposal which, in the short term at least, would terminate spent fuel transport from power stations north of the border – a change which many critics have advocated for all spent fuel.[16]

Nuclear Electric were looking at other options as well, including the extension of on-site storage facilities at AGR power stations, which would defer the need to transport fuel off site. They were even reported to be considering the possibility of sending fuel to France for reprocessing by Cogema at La Hague! This would not only redirect current oxide fuel movements from the AGR power stations – depending which sea crossing was used – but could also involve moving AGR fuel currently stored at Sellafield.[17]

7   Spent fuel – imports

Imports of spent fuel from nuclear power stations overseas have been arriving at Sellafield for almost as long as domestic material. Between 1965 and 1988, some 2,050 tonnes of irradiated magnox fuel and 1,900 tonnes of irradiated oxide fuel were imported into Britain.[1] Regular deliveries commenced in January 1966 with the first flasks from the Latina reactor in Italy. Three years later, shipments began from the Tokai Mura power station in Japan. Ever since, both Magnox power stations – the only two ever exported by Britain – have returned their spent fuel to Sellafield for reprocessing.

In the late 1960s, contracts were also won to reprocess oxide fuel from foreign light water reactors (LWRs). These reactors – either Pressurised Water Reactors (PWRs) or Boiling Water Reactors (BWRs) – consume low-enriched uranium oxide fuel. The first consignments arrived in Britain in 1968 from the Garigliano reactor in Italy.

Most, however, has arrived from Japan. Imports expanded particularly after 1973, when deliveries from Tokai Mura were augmented by increasing quantities of fuel from Japanese PWRs. Ironically, that year had another significance for BNFL: 35 workers were contaminated in an accident at Windscale's Head End Plant. This was a part of the Magnox reprocessing plant which had been converted to accept oxide fuel for reprocessing. Opened in 1969, it managed to process some 90 tonnes of oxide fuel before being closed by the accident. After this setback BNFL applied for planning permission for a new reprocessing plant known as THORP – the Thermal Oxide Reprocessing Plant.

The application led to the Windscale Inquiry where, for the first time in Britain, the whole concept of reprocessing and its political and social consequences came under critical review: the economics, the disposal of radioactive waste, the pollution of the Irish Sea, and the national and international military implications of extracting large amounts of plutonium. THORP was approved by parliament in 1978 and although the Inquiry is now history, the issues raised have become more relevant with the passage of time: THORP, which is now built, has had a major effect on the transport of radioactive materials generating more imports of irradiated fuel, with a promise of uranium, plutonium and radioactive waste exports in return.

Spent fuel imports, especially from Japan, have been both a national and local issue. It was news of the Japanese contracts in 1975 which led to the "nuclear dustbin" tag. By 1991, the total amount of Japanese fuel to be reprocessed – 2,283 tonnes – exceeded even the combined total of oxide fuel from Britain (from the Advanced Gas-cooled Reactors).[2] Most of these imports have arrived by sea at Barrow-in-Furness, first through the shipyards at Vickers then at BNFL's own berths in Buccleuch and then Ramsden Docks. In Barrow, the THORP proposal and the prospect of a massive expansion in nuclear trade, aroused local opposition with health and safety issues to the fore. With impeccable timing, shortly after the Japanese contracts were publicised in 1975, a train carrying 22 tonnes of Japanese spent fuel from the docks to Sellafield derailed near Barrow town centre. Although only the engine left the track, it was a timely reminder of the potential dangers to the town.

Health issues have been raised on a more personal level. The Barrow-based pressure group CORE (Cumbrians Opposed to a Radioactive Environment) has reported the case of Glenys, who worked in the Vickers shipyards when it was used for unloading imported flasks: ". . . her father remembers that on cold winter mornings she would lean up against the warm nuclear waste flasks as they were unloaded in the shipyard."[3] Warm because flasks of spent fuel radiate heat from the deacy of their radioactive contents. Also given off but neither seen nor felt by Glenys were small doses of gamma radiation, which penetrate even the thickest steel sides of a flask and certainly a human being. In 1977, Glenys was diagnosed as having breast cancer; four years later cancer of the brain was found. At the age of 31, Glenys died. Of course, it might not have been caused by spent fuel; she had been exposed to radiation from other sources as well. Employed as a cleaner in the shipyards, Glenys handled contaminated laundry; as a child she had played on local beaches and swum in Walney Channel, the contaminated waterway separating Barrow and its shipyards from Walney Island where Glenys had lived. Indeed her death may not have been caused by radiation. As the industry well knows, the cause of such cancers cannot be proved – one can only speculate. Nevertheless, considering Barrow's long involvement with the nuclear industry and spent fuel transport, speculation suggests that cancers in the same place are more than coincidence.

Just as the Japanese contracts have underpinned the economics of THORP, so they have guaranteed a steady flow of radioactive imports through Barrow. Statistics show the extent of this traffic. Deliveries from Tokai Mura have been running at about 50 tonnes per year.[4] Imports of Japanese oxide fuel have also built up with about 150 tonnes arriving per year.[5] For Barrovians opposed to this traffic, the fact that most of it cannot be reprocessed until THORP comes into operation has made the associated dangers even less acceptable.

Imports from Japan

Since 1976, spent fuel from Japan has been transported to Europe by Pacific Nuclear Transport Ltd (PNTL), a subsidiary company of BNFL and French and Japanese companies including three of the electricity utilities with reprocessing contracts at Sellafield – Tokyo Electric Power, Kansai Electric Power and the Japan Atomic Power Company.

PNTL owns its own fleet of purpose-built ships which also deliver Japanese spent fuel to the French reprocessing plant at La Hague. Three of the ships – the *Pacific Crane*, the *Pacific Teal* and the *Pacific Pintail* – have a capacity of 24 Excellox flasks, while the *Pacific Swan* and *Pacific Sandpiper* can carry up to 20 Excellox flasks with an additional 4 and 8 Magnox flasks respectively. The vessels are operated by James Fishers and Sons (Barrow-in-Furness) who act as managers and agents on behalf of PNTL.

The non-stop journey from Japan to Barrow and Cherbourg (for La Hague) takes PNTL's ships and their radioactive cargo through the Panama Canal and across the Pacific and Atlantic Oceans, a voyage of approximately 44 days. By 1988, the fleet had completed 50 round trips between Europe and Japan. After being unloaded at BNFL's berth at Barrow, flasks continue their journey to Sellafield by rail on purpose-built 16-wheeled railway wagons.

Imports from Europe

Most European imports have been handled by the Risley-based company Nuclear Transport Ltd (NTL). Since 1973, they have been responsible for the movement of all European oxide fuel to both Sellafield and the French reprocessing plant at La Hague. Latina remains the one exception; spent fuel from this power station is moved by BNFL itself. Flasks are loaded onto BNFL's ship the *Mediterranean Shearwater* at Anzio for shipment to Barrow. With the closure of the reactor in 1987, imports from Latina are coming to an end.

NTL are, in effect, the European counterpart to Pacific Nuclear Transport Ltd. although imports of European spent fuel have been far less frequent than those from Japan. Nevertheless, between 1974 and 1985, NTL imported over 500 tonnes of spent fuel from Canada, Italy, the Netherlands, Spain, Sweden, Switzerland and West Germany.[6]

Most deliveries arrive from Germany. BNFL's reprocessing contracts have covered fuel from over half of the country's existing nuclear power stations. Sellafield also receives spent fuel from the Beznau and Goesgen nuclear power stations in Switzerland and from the Dodewaard reactor in the Netherlands. Flasks are transported by rail across Europe and Britain on NTL's fleet of purpose-built 16-wheeled wagons.

For many years, flask wagons travelled through Belgium to Zeebrugge crossing the Channel on Sealink's train ferry to Harwich. Since the withdrawal of this ferry in 1987, flasks arrive on the train ferry service between Dover and Dunkerque, operated by SNCF's *Nord Pas de Calais*, now the only train ferry between Britain and mainland Europe.

The route from Germany may change in the future; an Essen-based company involved in the transport of radioactive materials, Gesellschaft fur Nuklear-Service (GNS), has expressed interest in building a rail ferry for making German deliveries to Sellafield.[7]

Imported flasks – imported controversy?

Compared to the fuss made over the years about the movement of domestic spent fuel, imports of similar material have generally attracted less attention. Yet although the amount of imported traffic is less, the issues raised are just as contentious, if not more so.

Foreign imports are why Sellafield has been called "the world's nuclear dustbin". Imports of "nuclear waste" have met objections as a matter of principle. BNFL, of course, see it differently. In a paper presented to the 1987 Uranium Institute annual symposium, BNFL's Chair Christopher Harding complained about anti-nuclear organisations using the "nuclear dustbin" tag "despite the provisions in our contracts for returning to the country of origin the wastes arising from the reprocessing of overseas fuel".[8] Yet this is not wholly true, as Harding himself acknowledged two paragraphs later. Referring to the Irish sea, he noted that part of its radioactivity came "from the discharge of low-level radioactivity in liquid form from Sellafield" – discharges which result from reprocessing, including imports from overseas. The famous provisions for returning waste are not entirely what they are made to appear. BNFL policy is not to return all of the different types of wastes arising, but only an equivalent amount of radioactivity in the form of high level waste. Thus while relatively small volumes of high level waste will eventually be returned to the customer, the low and intermediate level material – like the contamination in the Irish Sea and on the Cumbrian coast – will remain behind.

In any case, no such provisions were made in reprocessing contracts signed before 1976. When that material is reprocessed – after THORP comes on stream - the resulting wastes will have to be disposed of in Britain, namely some 100 cubic meters of high level waste, 3,000 cubic meters of intermediate level wastes and 20,000 cubic meters of low level wastes.[9]

At the ports of entry, imports of foreign spent fuel have provoked more local concerns. The transfer of flasks from ship to land raises several questions of health and safety. At Barrow, flasks are unloaded in a port where nuclear submarines are built and hazardous chemicals and fuel are stored. The dangers of such a concentration of industrial hazards were raised when BNFL applied for permission to built their new berth in Ramsden Dock – opposite a site used for the storage of gas condensate. Although objectors failed to stop the development, BNFL's new facilities, used only for radioactive material, have incorporated special security, surveillance and safety features, including tall columns of fire fighting equipment.

Such features are conspicuously absent at the other ports. At Dover's Western Docks spent fuel is treated like any other consignment of freight; flasks are shunted off the train ferry while coupled to other wagons which may include hazardous or inflamable chemicals. Incidents such as the diversion of the train ferry services to Dover Eastern Docks in September 1990 after explosive fumes leaked from a railway wagon in the Western Docks, highlight the potential dangers of not segregating hazardous materials.[10]

At Immingham, where spent fuel arrived from the Netherlands until 1988, the contrast with Barrow was no less acute. Irradiated fuel (in an NTL-15 flask on a lorry) arrived in the port on a roll-on roll-off ferry service from Rotterdam. On arrival, the flask was driven to a road/rail transfer facility near to the port to be loaded on to an NTL rail wagon for the rest of the journey to Sellafield. These arrangements caused a local outcry in 1988. Councillors questioned the wisdom of carrying out the transfer on an industrial estate near a residential area. Trade unionists expressed concern that the movement of radioactive materials through the port might discourage industry – especially food companies – from moving to the area and adversely affect the local ecomony. The arguments culminated in a decision by dock workers to refuse to handle irradiated fuel.

The contrast between operations at BNFL's berth in Barrow and the other ports where flasks have arrived is mirrored in the design of the ships used to transport spent fuel. BNFL's own vessels incorporate various special features to improve safety and security. These include "strengthening of the hull against collision impacts; watertight subdivisions to give adequate reserve buoyancy; additional fire fighting equipment of a particularly high standard; extensive duplication of machinery, equipment for navigation, communications and ship location (including satellite navigation equipment), cargo monitoring aids and cooling equipment in the holds."[11]

Such features are in marked contrast to the design of cross-Channel ferries, a subject of much debate especially since the capsize of the *Herald of Free Enterprise*. Although that ship did not carry railway wagons, train ferries have also had their share of misfortune. In 1982, for example, six people died when the Sealink train ferry *Speedlink Vanguard* collided near Harwich with Townsend Thoresen's *European Gateway*.[12] At Dover, where spent fuel has arrived since the Harwich service was withdrawn, two Sealink train ferries – the *Saint Eloi* and the *Cambridge Ferry* – crashed head-on in poor visibility in 1987.[13] Cross-Channel ferries may or may not be safe but they make no special provision for the transport of radioactive material.

The movement of flasks from port to Sellafield presents a similar contrast. While Japanese imports arriving at Barrow continue to Sellafield on dedicated trains (like flasks from British nuclear power stations) their European counterparts are conveyed on regular freight services and marshalled next to other wagons.

An Excellox flask for transporting Japanese spent fuel parked outside Sellafield. Background: BNFL's THORP reprocessing plant.

The contrasting ways in which flasks are transported at sea and handled at ports of entry suggests conflicting attitudes to safety. One of the underlying principles of spent fuel transport is that safety is assured by the design of the flasks alone, regardless of transport mode or other transport arrangements.

This is the view of the Department of Transport, who note that, "There are no regulations either international or national which require ships carrying spent nuclear fuels to be specially constructed or equipped".[14] If that then is the case, why the need for extra security and safety measures in BNFL's ships and at their docking facilities in Barrow? If the extra precautions at Barrow are because of other nearby hazards, then should fuel be imported though the port at all? Conversely, if BNFL's operations are taken as a model, then the routine shipment of spent fuel through other British ports looks comparatively lax.

Such questions are increasingly relevant because imports of spent fuel are set to increase, especially from western Europe. During 1990, BNFL announced several new reprocessing contracts with electricity utilities in Germany. The contracts involve reprocessing almost 1,600 tonnes of spent fuel from seven nuclear power stations: Gundremmingen-B and -C, Emsland, Isar-1, Philippsburg-2, Kruemmel, and Muelheim-Kaerlich. The contracts guarantee that about 45 per cent of West German utilities' spent fuel discharged to the year 2005 will cross the Channel to Britain.[15] (It is worth noting that the German contracts contain a political *force majeure*: if a future German government decided that spent fuel should be stored for final disposal without being reprocessed, the utilities reprocessing contracts could be cancelled – imports of spent fuel would then have to be returned to Germany.)

Despite its lower profile, the transport of imported spent fuel has not been without problems. Nor is it immune to some of the criticisms raised about the use of domestic flasks. After the CEGB's Old Dalby train crash, for example, critics pointed out that whatever conclusions might be drawn from that demonstration, they could only be applied to that type of magnox flask – also moving around on Britain's railways were flasks of different designs which were not tested at Old Dalby: some, like the British A1 AGR flask, might look similar but have much thinner sides; others, like those used to move imported spent fuel, are totally different. Although the survival of a magnox flask in a 100-mph train crash no doubt reassured many people concerned with domestic traffic, what would be the effects of such a crash on different flasks carrying different types of fuel?

As far as flasks used to transport imported spent fuel are concerned, the answer can only be estimated because full-scale tests have never been carried out. Like the British AGR flasks, tests to comply with the IAEA regulations have only been performed on models. Predictions about flask behaviour in accidents are made by extrapolating the results of scale-model testing to full size – results which are not always viewed with confidence by the industry's critics. Indeed, BNFL make it clear that for some tests not even scale models may be necessary: "The most usual method of complying with design criteria in the UK is the use of scale models for impact tests, while thermal performance assessments (fire testing) are made through detailed calculations."[16]

Criticism of this approach was implied in the 1986 inspector's report on the Sizewell B Inquiry. Although the exact type of flask to be used was not specified by the CEGB, it would presumably be an Excellox model as used at present for imports. Commenting on the safety aspects of spent fuel transport, the inspector, Frank Layfield, noted:

> . . . I think it is highly desirable that:
> a) a *full scale* flask of a type intended to be used to carry spent fuel from Sizewell B should be subjected to sequential drop and fire tests, as specified by the International Atomic Energy Agency (IAEA), on an experimental basis before being so used; [author's italics]
> b) the IAEA should be consulted with a view to including a side-on impact with a rigid sharp edge in the tests to be satisfied under the UK regulations for the safety of spent fuel flasks.[17]

Cylindrical flasks have, however, been subject to full-size tests in other countries. In the US, for example, Sandia Laboratories performed a series of well publicised tests on 20 and 25 tonne flasks. Four spectacular crashes were staged in 1977 and 1978:

1 A flask carried on a lorry and trailer unit was propelled at 60 mph on a rocket-powered sled into a thick concrete wall.
2 An 84 mph repeat of Test No. 1 (using the same flask).
3 An 120-ton railway engine was crashed at 81 mph into another stationary flask mounted on a lorry and trailer unit, simulating the consequences of a flask stalling on a level crossing. (This is not a hypothetical scenario: one of the worst rail accidents in Britain occurred on 6 January 1968 when a 75-mph passenger train crashed into a lorry carrying a 120-ton elecricity transformer stalled on a level crossing at Hixton, in Staffordshire.)
4 A flask mounted on a railway wagon was crashed into a concrete wall at 81 mph and subsequently engulfed in a 125 minute fire of JP-4 fuel burning between 980°C and 1150°C.

The tests were carried out, amongst other reasons, to compare predictions previously derived from models with the results of a test on the real thing. The tests, which were filmed for Sandia Laboratories, are cited by BNFL as evidence of the integrity of the flasks which bring imported fuel into Britain. However, whether tests on flasks of a different design can substitute for full-sized tests on NTL and Excellox flasks is debatable. For example, the US flasks used in the Sandia tests were only about one third of the weight of those used in Britain.

In any case, the Sandia tests themselves have been criticised. The flasks used were apparently obsolete, having been withdrawn from service in 1967. They were pressurised to only 26 psi, while over 130 psi might be maintained in actual use. In at least one of the tests, 100 cc of water leaked from the flask, although the film's soundtrack claimed that if radioactive material had been inside (which it wasn't) radiation would not have been released. The film also emphasised that the flask in the fire test was undamaged during a 90-minute period, but failed to mention that ten minutes later its outer shell cracked open in two places, the shielding began to vaporise and the test had to be stopped. The cracking was apparently due to a manufacturing and welding fault.[18]

On the other hand, the US tests were in some ways more rigorous than those carried out in Britain. Not only was the impacted flask in the final test subsequently subjected to a fire – a full-scale compliance with the IAEA regulations regarding cumulative impact and fire testing – but a higher temperature was sustained for a longer period than the IAEA requirements adopted in British fire tests.

Second, crashing a flask into a concrete wall seems like the sort of test that the CEGB should have performed after reading the report of their consultant's, Ove Arup and Partners. This report, referred to in the previous chapter, concluded that an impact against an unyielding surface (such as a bridge or tunnel abutment) was potentially the most damaging accident that a flask could experience – more damaging, in fact, than being struck by a high speed train.

Third, the first Sandia test used a flask containing 216 kgs of uranium. Not spent fuel, but an unused and unirradiated fuel element. The failure to test the effects of an impact on spent fuel was another criticism of the Old Dalby test: broken fuel would increase the release of radioactivity if a flask was also breached.

Fuel breakage is a potentially greater hazard with spent fuel from reactors such as the AGR or the PWR whose fuel elements or fuel pins contain uranium oxide in pellet form. One aspect of this problem was highlighted by the Union of Concerned Scientists (UCS) in the USA who studied the possible release of radioactive caesium from spent fuel.[19] The UCS have described how caesium, a fission product, "migrates" within an irradiated fuel element during normal operating conditions in a reactor; as much as 20 per cent escapes from the uranium and collects in the gaps between the cladding encasing the fuel and the fuel itself. Assuming that an accident severe enough to breach a flask would also break some of the fuel elements, these reservoirs of caesium would be released within the flask and eventually into the environment, escaping either as a gas or dissolved in leaking cooling water – bearing in mind that caesium is but one of many radioactive gases, liquids and solids within a spent fuel element. Reasssuringly, the force of the impact in the Sandia test – some 20 g – produced no detectable damage to the fuel element.

While opposition to spent fuel imports in the past has been mostly confined to the ports of entry, the current increase in flask traffic is beginning to attract a wider audience. For example, local councils along the routes are already taking a critical interest. The transport of foreign spent fuel looks set to repeat the controversies previously reserved for domestic traffic.

Box 7.1: Flasks used for imported spent fuel

Japanese spent fuel is transported in Excellox flasks. By 1988, PNTL were operating a flask pool of 80, made up of three different models. The most recent are 26 Excellox 4 flasks with a capacity of 7 PWR fuel assemblies. The remainder comprise 3 Excellox 3 flasks and 51 Excellox 3B flask. The Excellox 3, the original design for Japanese fuel, was introduced in 1973 and has a capacity of 14 BWR fuel assemblies. The Excellox 3B model is a slightly longer version of the Excellox 3 and carries either 5 PWR or 14 BWR fuel assemblies.

All Excellox models are cylindrical, typically about 2 m in diameter and approximately 6 m long. Flasks are made out of 90 mm carbon steel plates, rolled and welded into a single tube. One end accommodates a 300 mm thick steel lid - bolted and sealed into place after loading – while the other is plugged with a 350 mm thick base. Both the lid and the base are protected by balsa wood shock absorbers encased in stainless steel. Inside the steel body is a 160 mm thick lead liner which provides gamma and neutron shielding (as in the AGR flask). In the Excellox 4, the lining is 190 mm thick, and in all models, extends the full length of the flask cavity. Cooling fins and support legs are welded onto the outside.

Each flask contains a cylindrical "multi-element bottle" (MEB) which holds the fuel during transport; fuel elements are first placed in the MEB which in turn is loaded into the cavity of the flask. Irradiated oxide fuel, like any other, is intensely radioactive and continuously gives off heat. This is removed by filling the flask with water; by convection and conduction, heat is transferred from the fuel to the outside surface of the flask.

European imports to Sellafield are mostly transported in NTL flasks. NTL began life with a pool of just five cylindrical flasks previously owned by its parent companies, BNFL, Transnucleaire S.A. and Transnuklear GmbH. By 1988, the total had increased to 18. The company also operate a further 9 flasks belonging to Cogema.

Box 7.1 (continued)

Other flasks are also used on occasions. In 1989, for example, fuel from the German Unterweser nuclear plant was transported to Britain for the first time in a Castor flask. Developed by GNS of Essen, the Castor flask has, as its name suggests, a cast-iron cylindrical body with a stainless steel lid.

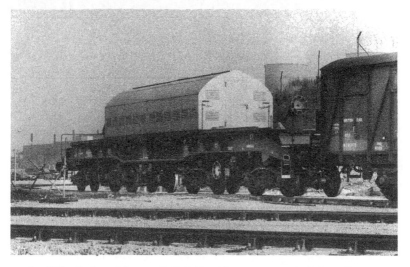

An NTL flask arriving at Sellafield with more spent fuel from Europe. NTL flasks are cylindrical flasks (like Excellox designs) although they are transported concealed under a metal housing.

A Castor flask transporting spent fuel from the Unterweser nuclear power station, Germany to Sellafield, on the *Nord Pas de Calais*, the train ferry from Dunkerque to Dover. The cylindrical flask is concealed under the metal housing.

8   Plutonium

The silvery white metal that destroyed Nagasaki was aptly named. Like the farthest known planet in our solar system, its name derives from Pluto, god of the underworld. As the most hated of Greek gods, this mythological demon offers some interesting parallels with its twentieth century reincarnation. Although Pluto's name came from the Greek word for wealth, it had an ambiguous meaning, paradoxically symbolising both good and evil. Beneficially, the underworld – if the word is taken literally to mean the earth beneath us – produced the crops which sustained the agricultural prosperity of ancient Greece and thereby life itself; conversely, the underworld was also the home of the dead where Pluto reigned as the heartless enemy of mortal life. Black sheep were offered as a sacrifice and the symbol of his infernal empire was a helmet which made him invisible.

These ancient myths have a worrying familiarity; Pluto's contradictory associations with wealth and death have been reborn in nuclear science. On the one hand, plutonium can be used to fuel reactors which, according to their proponents, promise cheap and plentiful electricity; conversely, it is the raw material for a nuclear holocaust. As for Pluto's invisible presence and his sacrificial sheep – were these not part of the legacy of Chernobyl after its poisonous clouds had rained over Britain? Pluto is indeed a god for our time.

Plutonium's popular reputation as one of the most deadly materials around is not of course shared by the industry. BNFL's Transport Manager, William McLaughlan, once astounded an audience of Scottish councillors, concerned about the transport of the material, by remarking that you could eat a lump of plutonium the size of an orange and it would not do you any harm. In fact, there is some radiological basis for this claim: plutonium emits mostly alpha particles, which penetrate less than 0.1 mm. of human tissue and will not pass through unbroken skin – arguably, ingested plutonium might pass through the body without causing too much damage.

However, disregarding the fact that it is also a toxic heavy metal (and few of those are considered safe to eat), it is disingenuous advice even on radiological grounds. Plutonium is most hazardous if *inhaled* – as a fine powder, dust or aerosol – or absorbed into the blood. Margaret Gowing (the UKAEA's official historian) painted a graphically different picture when she described how "more than one scientist was known to say that if he thought he had touched it with a cut finger he would immediately cut his finger off".[1]

Hardly surprising, then, that a vital material for nuclear weapons and one which is increasingly used as a nuclear fuel should attract controversy. Even back in 1976, a Royal Commission on the Environment noted: "We should not rely for energy supply on a process that produces such a hazardous substance as plutonium unless there is no reasonable alternative".[2] Yet despite that warning, plutonium is now in regular use producing electricity, principally in fast breeder reactors. In Britain, the development of fast reactors has been concentrated at Dounreay in Scotland, a UKAEA site which includes one working fast reactor, another now closed, and Britain's only fast reactor fuel reprocessing plant. With the civil and military uses of plutonium now well-established, the transport of plutonium in Britain can be summarised as follows:

1 Round trips between Sellafield, Dounreay and France of plutonium (in various forms) for use in the fast reactor programme.
2 Round trips between Sellafield and the nuclear weapons factories at Aldermaston and Burghfield of plutonium used in nuclear weapons.
3 Exports of plutonium from Sellafield to other countries for various civil and military uses.

The movement of civil plutonium

Most of the plutonium transported in Britain for civil purposes is used within the fast reactor fuel cycle. In Britain, the development of fast reactors has been concentrated at the UKAEA's site at Dounreay on the northern coast of Scotland – conspicuously about as remote from major centres of population as one can get in mainland Britain. The UKAEA's first large prototype fast reactor – the Dounreay fast reactor (DFR) – was commissioned here in 1959. In fact, the DFR, which operated until 1977, was fuelled with highly enriched uranium rather than plutonium. Only when the larger plutonium-fuelled Prototype Fast Reactor (PFR) began operating at the same site in 1974 did the movement of sizeable quantities of plutonium begin for civil use.

With a fast reactor and, since 1980, Britain's only fast reactor fuel reprocessing plant, Dounreay is almost self-contained – but with one vital function missing: it does not manufacture new plutonium fuel. This is carried out by BNFL at Sellafield, thereby generating movements of plutonium between the two sites and completing the fast reactor fuel cycle in Britain. The plutonium recovered from used fuel at Dounreay is extracted in the form of plutonium nitrate. This intensely toxic green liquid is transported to Sellafield in massive 20 tonne transport containers, design No. 1763. Each container (measuring approximately 3 x 3 x 3.5 metres) holds up to 250 litres of plutonium nitrate solution, containing some 100 kgs of plutonium.[3] Most of the journey from Caithness to Cumbria is by sea, sailing from Scrabster – Dounreay's nearest port – to Workington on the Cumbrian coast. The blue containers are shipped two at a time, each mounted on the back of an articulated low-loader.

The sea journey is entrusted to the *Kingsnorth Fisher*, a roll-on roll-off ship specially designed to carry large or heavy loads. Unlike other ro-ro ships with bow doors at each end (like the ill-fated *Herald of Free Enterprise*), the *Kingsnorth Fisher* is a conventional ships but with a ro-ro deck built on top. Owned and operated by the Barrow-based company, Fishers, its flexible design means that it can be loaded, if necessary, without dock craneage or other facilities. The transport of plutonium nitrate is infrequent and by 1988, only 21 deliveries had been made since shipments first began.[4] Empty flasks are taken back to Dounreay on the return journeys which are also used to move consignments of radioactive waste.

Undoubtedly it is a matter of policy to reveal as little as possible about the transport of plutonium nitrate. Yet while it is understandable that the arrival and departure times of sailings are not publicised, even information about the general risks of plutonium nitrate transport has been suppressed. When the shipments were first proposed, considerable concern was expressed about the risks of transporting an extremely hazardous material through some of the roughest and most unpredictable seas around Britain: what would happen if the ship or a flask were to sink? With up to 500 litres of plutonium nitrate on board, could it leak and what would be the consequences if it did? What would happen if a shipment was hijacked by "terrorists"? Such questions were officially considered before the shipments began. Security aspects, for example, were assessed by the Department of Energy although the outcome of their deliberations has never been made public. According to the Energy Secretary: "The Government are satisfied with (the security) aspects; it would not however, be in the public interest to publish the advice to the government on this matter."[5]

Nor is it in the public interest, according to the government, to publish the safety reports on the transport of plutonium nitrate. These documents were prepared for the Health and Safety Executive by BNFL and the UKAEA, but none of the contents have been revealed because they include classified information. In fact, some of the conclusions have since surfaced in technical literature published by the industry. A 1980 conference paper, for example, referred to the consequences of engulfing a plutonium nitrate container in a fire. This is a remote but hardly inconceivable possibility either at sea or especially during the road journeys, when flasks are exposed to potential traffic accidents and the everyday dangers of life on land. The results would no doubt have interested the critics when shipments were first proposed: "It was estimated that the pressure vessel holding the plutonium nitrate would fail as its temperature approached 300°C. Calculations suggest that the temperature of the contents of the package would not rise significantly for the first two hours of any fire then the temperature would rise fairly steadily and reach the critical value of 300°C in 5 hours . . ."[6] However, as far as the public and their elected representatives are concerned, the suppression of the original reports suggests that secrecy is more important than safety.

Once at Sellafield, plutonium nitrate is processed into plutonium dioxide from which new fast reactor fuel is produced. The core of the PFR holds almost 80 fuel "subassemblies" most of which each contain 325 fuel pins made from plutonium and depleted uranium. The depleted uranium components are manufactured at Springfields, while the final assembly of all PFR fuel assemblies is carried out at Sellafield.

Until 1988, Sellafield was also responsible for manufacturing the plutonium fuel pins. Since then however, these have been made in France by the Commissariat à l'Energie Atomique (CEA) at their fuel fabrication plant at Cadarache, about 50 miles north of Marseilles. The switch to France was part of the collaborative arrangement between Britain, France, Belgium, Italy and West Germany to design and construct fast reactors in Europe. Manufacturing fast reactor fuel in Cadarache avoided a duplication of effort. The transfer of this part of the fast reactor fuel cycle to France has altered plutonium transport patterns and increased the overall number of plutonium shipments associated with the PFR: while Sellafield continues to process plutonium nitrate, its output – plutonium dioxide – must now be transported to France for the next stage of PFR fuel production. The French-made fuel pins are returned to Sellafield to be fitted into subassemblies for delivery to Dounreay. Diagramatically, Britain's fast reactor fuel cycle can be illustrated as follows:

Dounreay Prototype Fast Reactor (UKAEA)
↓
plutonium nitrate from reprocessed PFR fuel
↓
Sellafield (BNFL)
↓
plutonium oxide
↓
Cadarache (CEA)

plutonium fuel pins
↓
Sellafield (BNFL)
↓
new plutonium fuel subassemblies
↓
Dounreay Prototype Fast Reactor (UKAEA)

New fuel assemblies from Sellafield's PFR fuel line are periodically delivered to Dounreay. Until 1977, new fuel travelled to Scotland by road in guarded bullion vans.[7] Despite this security, the intrinsic vulnerability of road transport did not go unnoticed. In May 1977 – just months before the dangers of the "plutonium economy" became a central issue at the Windscale Inquiry – the *Daily Express* announced that it had commissioned the construction of a 1.3 kiloton atom bomb to demonstrate how easy it was to build a plutonium warhead.[8] According to the newspaper, the plutonium could be obtained by hijacking a road delivery of fast reactor fuel between Sellafield and Dounreay. The *Express* even printed a photograph of an isolated moor near the Scottish coast where the vehicles could be stopped. Not far from a small harbour, it offered potential bomb builders an ideal location to hijack the fuel and a convenient escape route from the country.

The UKAEA flatly refused to discuss the security of PFR fuel deliveries, although it was probably no coincidence that in the following January fuel was dispatched to Dounreay by air for the first time.[9] By 1987, new fuel was being delivered about 10 times a year. Most consignments are flown to the UKAEA's own airstrip within the Dounreay site, although deliveries have also been made via Wick airport, about 30 miles away.

While the air transport of plutonium has security advantages for the nuclear industry, it imposes greater risks on the general population in the event of a crash: the consequences of dropping to the ground from a high-flying aircraft are a bit more serious than falling off the back of a lorry. In any case, Sellafield itself has no airport and consignments of plutonium fuel must still travel some distance by road in order to get to the plane. Until 1979, fast reactor fuel was flown to Scotland from the council-owned airport at Carlisle, approximately 55 miles from Sellafield. However, flights from Carlisle came to an abrupt end when local councillors found out and immediately banned them.

For several years afterwards, deliveries to Dounreay departed from Speke airport on the outskirts of Liverpool. Unfortunately for BNFL, the movement of plutonium became an issue during the 1986 public inquiry into the European Demonstration Fast Reactor Fuel Reprocessing Plant (EDRP). In the previous year, BNFL and the UKAEA had submitted a joint planning application to build a new reprocessing plant at Dounreay to handle spent fuel from planned fast reactors in Europe. The project, which would generate a substantial increase in plutonium shipments, was eventually approved in 1989 although it looks unlikely to be built. When the Inquiry looked at existing transport arrangements, the discovery that plutonium fuel passed through Speke led to a predictable outcry on Merseyside. Both Liverpool City Council and Merseyside Metropolitan Council had declared themselves nuclear-free zones and in June 1986, BNFL were once again told to leave. Plutonium flights to Dounreay resumed from Carlisle where the local council, mindful of the extra income for its ailing airport, this time chose to keep quiet. Deliveries from Carlisle to Dounreay were made by Dan Air, using HS.748 aircraft, the largest capable of landing at Dounreay.[10] Although the exact route to Scotland is determined by Air Traffic Control, BNFL advise Dan Air to avoid passing over urban areas and, where possible, to fly over the sea.[11]

The transport of plutonium fast reactor fuel requires stringent precautions. It is highly radiotoxic and potentially vulnerable to what the industry calls "misappropriation". Transport containers for fast reactor fuel are considerably more substantial than those used for most other fuels: they must be proof against theft and designed to prevent the contents going critical. Consequently, New PFR fuel is transported in 3 tonne containers designated the "1356". Each container holds a maximum of just four fuel subassemblies and weighs about four tonnes when loaded. Four tonnes has also been the maximum payload of aircraft using the Dounreay airstrip, thus restricting air deliveries to one container per flight. A similar type of container is used for returning plutonium fuel pins from France to Sellafield. In addition to movements of new fuel, Dounreay also receives occasional air deliveries of plutonium residues from the PFR fuel production plant at Sellafield.

The movement of military plutonium

Plutonium for nuclear weapons is obtained from two sources; either new supplies are produced from irradiated nuclear fuel or old plutonium is recovered from obsolete warheads and recycled for use in new ones. Either way, the material passes through Sellafield, resulting in movements of plutonium between Cumbria and the warhead plants at Aldermaston and Burghfield.

Fresh supplies of plutonium for military use are manufactured as needed by British Nuclear Fuels in their reactors at Calder Hall and Chapelcross. The plutonium is extracted at Sellafield and delivered to Aldermaston. Recycling has been carried out since the earliest days of Britain's nuclear weapons programme.[12] Later generations of warheads incorporate nuclear materials previously used in earlier bombs. Sellafield's role in this aspect of the weapons programme has never been officially explained although a 1982 statement by the Defence Secretary provides a clue: "Obsolete nuclear weapons . . . are carefully dismantled and all reusable materials, including all special nuclear materials, are recovered and reprocessed."[13] Reprocessing is necessary because plutonium "decays' over a period of time. A growing percentage of it changes into other substances, including americium-241. Increasing amounts of this unwanted isotope increase the radioactivity of a warhead which can affect its potential yield. Thus ageing plutonium must eventually be purified to remove the americium.

The existence of a plant at Sellafield for purifying military plutonium was alluded to at the 1977 Windscale Inquiry. When BNFL were asked about the increasing discharges of americium-241 from the site during the 1970s, it transpired that they came not from the company's uranium fuel reprocessing operations, as might be expected, but from "other activities" which BNFL were reluctant to discuss. A written submission to the Inquiry from BNFL revealed that an annex to building B204 at Windscale was used for the recovery of "aged plutonium residues'.[14] In fact, Sellafield's plutonium link with Aldermaston had been identified many years earlier when the UKAEA, then responsible for the site, described its activities to a House of Commons Select Committee noting: "There is also a small metal recovery plant in which plutonium metal residues returned from Atomic Weapons Research Establishment (Aldermaston) are dissolved and recycled through the metal production line".[15] Such operations have generated a two-way traffic of plutonium between Berkshire and Sellafield. The material for recycling is transported to Sellafield from Burghfield; in the other direction, billets of fissile material are returned to Aldermaston.

The security and safe delivery of what the MoD call "SNM" (Special Nuclear Material) is one of the responsibilities of the MoD police at Aldermaston. Consignments of SNM – which includes highly enriched uranium (and possibly tritium as well) – are transported in a large blue Seddon Atkinson van of which two examples are known to exist, a four wheel and six wheel version. SNM consignments are accompanied by a Special Escort Group comprising armed MoD police officers from Aldermaston. These movements first attracted attention in 1982 when a peace camp was set up outside Burghfield. Camp residents noticed the arrival and departure of a "very ordinary looking" box van which despite its humble appearance was usually flanked by a pair of Land Rovers bristling with aerials and carrying police. The short convoy – which has latterly favoured Range Rovers as escort vehicles – would be preceeded by a civilian police car and followed by a minibus or Land Rover. At least one transport accident is known to have occurred. In March 1986, at about 4 o'clock on a Wednesday afternoon, a private motorist was involved in a traffic accident with the convoy – travelling from Sellafield – on the M4 between Newbury and Reading.

In future, however, movements of plutonium between Berkshire and Sellafield will decline. Although supplies of new material will still be delivered to Aldermaston, the transport of recycled military plutonium between the two sites should cease in the 1990s. The A90 complex being built at Aldermaston for the Trident programme reportedly includes facilities for dismantling obsolete warheads and a chemical reprocessing plant for "cleaning" the recovered plutonium.[16] This development will eventually eliminate the need to transport recycled plutonium offsite.

Imports and exports

Plutonium movements within Britain are augmented by an irregular traffic of imports and exports. These international shipments have been accounted for by three reasons in particular:

1 Traffic between Sellafield and France for Britain's fast reactor programme.
2 The UK/US Mutual Defence Agreements have enabled plutonium from British reactors to be exported to the US, partly in exchange for imports of US-produced highly-enriched uranium.
3 BNFL's contracts to reprocess foreign spent fuel at Sellafield result in a steady output of plutonium some of which is exported for BNFL's foreign customers.

Imports and exports of plutonium are no less secret than domestic shipments. Transport arrangements are unannounced and few details are released about the quantities involved – indeed, the total amount of plutonium exported to date has never been officially revealed. However, from the few figures which have been published a rough indication of the quantities involved can be gleaned.

The shipments between Britain and France as part of the fabrication of PFR fuel began in 1988. In 1988–89, exports of separated plutonium totalled 368 kgs which, apart from several samples for safeguards purposes (to Austria and Italy), went entirely to France. Plutonium imports from France during the same year amounted to 181 kgs, presumably in the form of fabricated fuel pins from Cadarache.[17] Deliveries to and from France are clearly erratic as figures for the following year show a very different picture. Exports totalled 1,148 kgs, and also included a "contract" consignment to Belgium and more samples to Austria, Belgium and Germany; as the quantities exported to these countries are not separately given, the amount sent to France in that year cannot be determined. Surprisingly, imports from

France for the same year were zero.[18]

Other imports and exports reflect the scale of BNFL's reprocessing operations at Sellafield. The infamous reprocessing plant is one of the largest extracters of plutonium in the world, with a capacity well in excess of British needs. In earlier years, this politically opportune surplus enabled Britain to barter domestically-produced plutonium for US supplies of highly-enriched uranium, an arrangement provided for by the 1959 Mutual Defence Agreement. Despite government assertions that the plutonium exported has only been used for civil purposes, the quantities involved have never been revealed. The official reason is always the same: "Because of the barter arrangements under which plutonium was consigned it would not be in the national interest to publish the figure."

However, unofficial estimates suggest that between 1964 and 1971 approximately 4 tonnes of plutonium from Britain's Magnox reactors were transported to the United States.[19] Most of this material was produced in reactors belonging to either the CEGB or the SSEB which were brought on stream in the second half of the sixties. The transport of plutonium to the US under the exchange agreements has continued since 1971, although the quantities exported have been far smaller and deliveries intermittent. (Later deliveries have originated not from the "civil" Magnox power stations, but from Britain's military stockpile.)

In contrast, civil exports, mostly to countries other than the US, have been increasing. These are the products of Sellafield's reprocessing contracts with overseas customers. So far, almost all of the reprocessing carried out by BNFL for foreign customers has been limited to magnox fuel from the two British reactors exported during the sixties: the Latina power station in Italy and Tokai Mura in Japan. Imports of oxide fuel must await the completion of THORP before they can be reprocessed.

Between 1965 and 1989, some 3.5 tonnes of plutonium, recovered from imported spent fuel, were exported from Sellafield, either to the country of origin or to another country specified by the owner of the plutonium.[20] However, with the exception of 50 kgs of plutonium exported to the US for civil purposes (between 1971 and 1981), the amounts exported to individual countries have never been officially disclosed. What few details have been released show only total exports over a period of years. For example, in 1981 the government revealed that 1.93 tonnes of plutonium derived from reprocessed fuel belonging to overseas customers had been returned (in consignments larger than gram quantities) during the previous ten years to Belgium, Canada, France, Italy, Japan, the United States and West Germany.

During the same period, a further 1.28 tonnes of plutonium produced in British reactors were sold for civil purposes (in consignments larger than gram quantities) to Belgium, France, Japan, Switzerland, the United States

and West Germany.[21] By 1987, the list of countries receiving consignments
greater than gram quantities included Austria, the Netherlands and Sweden,
in addition to those above.[22]

Most exports have been for use in fast reactors. With an almost
embarrassing surplus of plutonium, Sellafield has been a well-established
source of supply, especially for experimental reactors in France and Japan.
A 1983 report for the US Department of Energy showed that to 1982, over
half a tonne of plutonium separated at Sellafield from Latina's spent fuel had
been used to fuel the Superphœnix fast reactor in France while Japanese
reactors (the heavy-water reactor at Fugen and the Joyo fast reactor) had
between them received one tonne of recycled plutonium from Sellafield,
separated from Tokai Mura's spent fuel.[23]

Although neither BNFL nor the British government will divulge
information about individual deliveries, a partial list of export consignments
from Sellafield to Japan has been published in Japan. Between 1970 and
1981, eight air deliveries and five sea shipments were used to return some
660 kgs of fissile plutonium (recovered from Tokai Mura's spent fuel) (see
Table 8.1).

Table 8.1: Plutonium deliveries from Britain to Japan[24]

Year	Number of times	Amount shipped	Transport mode
1970	1	25 kg	air
1972	4	110	air
1973	3	60	air
1975	1	65	sea
1978	1	40	sea
1979	1	105	sea
1980	1	190	sea
1981	1	65	sea

(Japan received a further 480 kgs of plutonium between 1965 and 1984 from Britain,
France, West Germany and the United States.)

In addition to exports of plutonium extracted by reprocessing, additional
quantities of plutonium are imported and exported for experimental
purposes. Most of these movements travel to or from the UKAEA
although the quantities involved are relatively small. Between 1979 and
1986, for example, the UKAEA received 230 kgs of plutonium from
Austria, Belgium, France, Japan, Netherlands, Norway, Switzerland, the
US and West Germany.[25] Nearly all this material arrived in the form of

fuel pins imported for research.

International collaborations on fast reactor research have led to further movements of plutonium. Under a joint UK/US agreement signed in 1979, small quantities of experimental US fast reactor fuel have been irradiated in the Prototype Fast Reactor at Dounreay. The first consignment, reported in 1979, consisted of US-made fuel pins (containing 15 kgs of plutonium), imported into Britain to be made into fuel assemblies.[26] After irradiation at Dounreay, the fuel returned to the US for further testing. Irradiated plutonium fuel pins have been sent by rail from Thurso station to Liverpool, for subsequent shipment to the United States.

Imports and exports of plutonium have also resulted from fast reactor collaboration between Britain and France. As part of a joint experiment to simulate the behaviour of large plutonium reactor cores, most of the fissile material in the UKAEA's ZEBRA reactor at Winfrith was transferred to a French research reactor at Cadarache at the end of 1987. The experiment – code-named CONRAD – involved exporting about one tonne of plutonium metal in the form of ZEBRA fuel plates. A similar large transfer might occur to help save the PFR. In an effort to stave of its closure, the UKAEA have been trying to obtain unused fuel from the SNR-300 fast reactor at Kalkar in Germany which although built was never allowed to operate. Acquisition of the reactor's fuel would save the UKAEA money and help prolong the life of the PFR.

By and large, however, imports of separated plutonium, usually for research and development, are relatively small. With its rapacious appetite for foreign reprocessing contracts, Britain's nuclear industry is first and foremost an exporter of plutonium, a business which will expand dramatically in the near future. Two developments in particular will, if completed, drasticly increase such movements:

The European Demonstration Reprocessing Plant (EDRP), Dounreay.

Approved in 1989 after the public inquiry, this will – if built – extract plutonium from spent fast reactor fuel imported from Europe. Its output, plutonium dioxide powder, will be flown from Dounreay's own airstrip to the fast reactor fuel fabrication plant at Cadarache in France. Using the largest aircraft that can presently land at Dounreay – the HS.748 – over 200 deliveries per year might need to be made. However, the flagging fortunes of the European fast reactor programme have made it unlikely that EDRP will actually be built.

The THORP reprocessing plant, Sellafield.

When this comes into operation in the early 1990s, large quantities of plutonium will be extracted from foreign spent fuel and exported for BNFL's overseas customers. Just under half are European utilities while the rest are Japanese. Figures given in a 1987 BNFL report assumed that after THORP had begun operations, a maximum of 3 tonnes a year would on average be returned to the company's customers. Assuming air transport with each flight transporting two of BNFL's new 1680 packages (which hold up to 48 kgs of plutonium), there would be 30 flights a year – 15 to Japan and 15 to Europe – each carrying 100 kgs of plutonium.[27] Flights to Europe were planned from Carlisle airport, with Prestwick preferred for Japan. However, because of problems developing a suitable air transport container (see following section), sea transport is now favoured for deliveries to Japan. Shipments are expected to sail from BNFL's berth at Barrow.

In fact, the quantities exported are likely to be higher. By 1991, THORP's reprocessing contracts for the period 1993-2002 had reportedly reached a total of 6,700 tonnes (of uranium), including 2,200 tonnes from Japanese customers and over 2,300 tonnes from customers in mainland Europe.[28] Assuming a plutonium content of approximately 1 per cent, this suggests that 45 tonnes will be exported over this period – 2.2 tonnes per year to Japan with a similar amount travelling to Europe. Plutonium exports will of course continue after 2002 as THORP wins further reprocessing contracts for its second decade of operation and beyond.

The transport implications are complicated by a possibility that Japan may decide to have its plutonium fabricated into mixed oxide fuel (MOX) fuel in Europe before being returned. MOX fuel is made from a mixture of plutonium and uranium but used in ordinary nuclear power stations. Its fissile content is obtained by substituting plutonium for enriched uranium. In the absence of any significant demand for fast reactor fuel, MOX fuel is an attractive way of utilising growing stocks of plutonium recovered by reprocessing spent fuel. It is increasingly being used in European reactors. (The French nuclear power station at Gravelines near Calais has been partly refuelled with MOX comprising 5 per cent plutonium and 95 per cent depleted uranium.) Small quantities of MOX fuel have also been manufactured over the years by BNFL at Sellafield, mainly for use in the Windscale AGR and the SGHWR at Winfrith.

Although BNFL have plans to begin operating a MOX fuel fabrication plant at Sellafield in the late 1990s, MOX fuel is not currently manufactured in quantity in Britain. If MOX fabrication was required before then, Japanese plutonium would have to be transported to fuel manufacturers in Belgium, France or Germany – thus increasing the number of plutonium movements within Europe. Fabricated MOX fuel would be shipped to Japan by sea, possibly from Britain after being returned from mainland Europe.

There are also exports of plutonium fuel samples and these too are set to increase. Samples weigh at most a few grams and consist of a powder of either plutonium dioxide or a mixture of plutonium and uranium oxides. Most are flown out of Heathrow and Ringway (Manchester), arriving at the airport by car. During the 1990s, the number of samples exported will increase to about 40 a year.[29]

Plutonium spreads its wings

With plenty of experience at moving plutonium, the nuclear industry has been remarkably successful at keeping it out of the public eye. With a few spectacular exceptions (principally outside Britain), most shipments to date have passed without attracting attention. Between the mid-70s and 1986, for example, BNFL dispatched 70 consignments of plutonium to overseas destinations – 90 per cent of which were delivered by air, and over 90 per cent of those in the form of plutonium dioxide.[30] Such deliveries – involving an average of 30 kgs of plutonium dioxide per shipment[31] – have been moved with a minimum of fuss; unobtrusive in the extreme. Plutonium consignments (in larger than gramme quantities) have been flown out of Carlisle, Dounreay, Wick and Hurn (Bournemouth) airports; separated plutonium in the form of samples and specimen fuel pins is also transported through the ports of Newhaven, Portsmouth and Liverpool.[32]

Outside Britain, however, the story has been different. In the United States in particular, plutonium transport has long been a controversial issue. One of the first manifestations of concern was a 1975 embargo by pilots belonging to the US Airline Pilots' Association (ALPA). To reduce the risks of hijacking and terrorism, ALPA pilots decided not to carry "strategic special nuclear material" on their planes – e.g. weapons-grade plutonium or highly-enriched uranium. Such substances, said ALPA, should be air transported only on military planes and through military bases. Although the protest was limited mainly to domestic routes, the pilots' action effectively prevented the air transportation of plutonium on all civil flights within the US. The inconvenience caused by the pilots was compounded by the US Transportation Safety Act, enacted by Congress in 1975. This legislation prohibited the transportation of all radioactive materials on passenger-carrying aircraft unless the materials were intended for research or medical use and were not an unreasonable hazard to health and safety.

This legislation was followed later that year by Public Law 94-79 which imposed a specific embargo on the transport of plutonium on domestic passenger aircraft within the United States. Public Law 94-79 prohibited the air transport of civil plutonium (with the exception of small quantities for research or medical use such as plutonium-powered heart pacemakers) until the US Nuclear Regulatory Commission could certify to Congress "that a safe container has been developed and tested which will not rupture under crash and blast-testing equivalent to the crash and explosion of a high-flying aircraft". Thus by protests and then by legislation, the air transport of virtually all civil plutonium within the United States was grounded.

Of course, plutonium did not stop moving around – it simply went by road instead. In any case, imports and exports continued to be flown in and out of the country as before: the difference being that plutonium had to be transported to and from international airports within the United States by road instead of air, as New York City discovered in 1975 – massive consignments of plutonium, imported "from somewhere in Europe" to the Westinghouse Plutonium Fuel Development Laboratory in Cheswick, Pennsylvania, were passing through Kennedy Airport. As if that wasn't enough, it transpired that the plutonium had been moved to Cheswick by road – across Manhattan, over the George Washington Bridge and through the state of New Jersey. Industry reassurances did little to placate the ensuing furore.

Worries about plutonium shipments could not have been eased by alleged violations of the transport regulations by Transnuclear, the company responsible for the New York deliveries. On one occasion, regulations limiting the in-transit storage time (between arrival and departure) of strategic special nuclear materials at an airport to a maximum of 24 hours, were inadvertantly broken when the scheduled flight was cancelled. On another, Transnuclear was fined $2,100 by the Nuclear Regulatory Commission (NRC) for not providing the obligatory armed guard during the transfer of an export consignment of enriched uranium hexafluoride from a domestic to an international flight.

With calls for a ban on international plutonium shipments through Kennedy Airport, concern also grew about the movement of other radioactive materials – especially spent fuel – through densely populated areas. On 15 January 1976, New York City responded to mounting pressure by banning the transport of most nuclear materials through its area. The potential dangers of plutonium were summed up at the time by Leonard Solon, Director of the City's Health Department Bureau for Radiation Control. Solon recalled the worldwide concern about fallout from atmospheric testing of nuclear weapons and the scientific pressures which had led to the Partial Test Ban Treaty of 1963. According to Solon, nuclear tests by the US, Soviet Union, Britain, France, China and India had released, among other dangerous substances, an estimated 400,000 curies of plutonium – a slightly lower total of radioactivity than the contents of a single plutonium shipment through Kennedy Airport. In the year preceeding the ban, six such consignments – each with a typical activity of half a million curies – had passed through Kennedy Airport. Solon concluded: "Dispersion of even a small fraction of the contents of one of these shipments as a result of an air crash, concomitant fire, and high winds within the City of New York . . . could have cataclysmic results bringing death or injury to thousands of New Yorkers."[33] New York City's ban was the beginning of a long legal dispute as the nuclear industry challenged the City's legislation. Although it remained in force for several years, the ban was eventually overturned by a Federal court.

The issues raised in New York surfaced on the other side of the Atlantic in 1984 when 189 kgs of fissile plutonium were shipped from France to Japan. This consignment, for eventual use in the Joyo fast reactor, had been recovered from Japanese spent fuel reprocessed at La Hague by Cogema, the French equivalent of BNFL. Because the plutonium had been extracted from fuel originally supplied to Japan by the United States, its return to Japan required US Congress approval under the 1978 US Nuclear Nonproliferation Act.

On this occasion, the plutonium (packed in 16 French FS-47 flasks and four British 1356 flasks) went by sea, sailing from Cherbourg on the *Seishin Maru*, a former Japanese ore freighter specially adapted for the job: security measures included independent communications systems linked to US satellites which allowed the ship's voyage to be monitored continuously. For part of the journey the vessel was accompanied by a US military escort unit to minimise response time in the event of an incident.

The shipment, made shortly after the *Mont Louis* sinking, received extensive publicity in Europe and Japan. With plenty of advance warning, it attracted hostile demonstrations at both ends of the journey. For the nuclear industry, it must have provided an invaluable lesson in how not to move plutonium – the disadvantages of marine transportation were embarrassingly obvious: Greenpeace attempted, unsuccessfully, to blockade the ship's departure from Cherbourg, while a front-page aerial photo of the ship approaching Tokyo gave Japanese demonstrators ample time to muster at the docks. In between, the ship's passage through the Panama Canal managed to provoke a diplomatic incident. The US-controlled Panama Canal Commission had previously announced that in accordance with international safety procedures, the ship would pass through the waterway in daylight. In the event, the *Seishin Maru* sailed at night reaching the Pacific Ocean at 2.32 a.m. on a Sunday morning, reportedly to avoid yet more anti-nuclear demonstrations. The change of plan to sail through the canal in the dark was condemned as dangerous by the Panama government which lodged a formal protest to the United States.

The security aspects of the *Seishin Maru* operation led the United States to decide that future deliveries of plutonium extracted from US fuel would have to be made by air instead. Air transport has long been favoured for security reasons: plutonium is harder to get at when it is off the ground. Moreover, a typical 12-hour flight from Europe to the Far East is a lot cheaper and easier to arrange than a comparable sea journey which can last at least six weeks. Unfortunately, air transport has not been a straightforward alternative. In the wake of Public Law 94-79 and the suspension of civil plutonium flights within the United States, the NRC devised more stringent criteria for testing the integrity and accident performance of plutonium transport containers. These were published in January 1978 as Nureg 0360: "Qualification Criteria to Certify a Package for Air Transport of Plutonium".

Nureg 0360 was not the obscure technical document it might sound. Because several of its tests were significantly more severe than those recommended by the IAEA, it has had a direct bearing on the transport of plutonium around the world. For example, while the IAEA impact tests represent the effects of a 30 mph crash, Nureg 0360 required subjecting a plutonium package to an impact of at least 288 mph onto an unyielding surface. Similarly, the fire test specified that a plutonium package must be exposed to a temperature of at least 1,010°C (from a pool fire of JP-4 or JP-5 aviation fuel) for a period of at least 60 minutes. Both the time and the temperature exceed the IAEA's traditional test criteria: 800°C for 30 minutes – figures quoted with divine authority whenever Britain's nuclear industry wants to reassure its transport critics.

More specifically, Nureg 0360 created a problem for BNFL: the plutonium containers which they had been using – notably the 2816C model – did not meet the new US criteria. (Plutonium has therefore been flying over Britain in containers which, by current US standards, have been inadequately tested.) Although Nureg 0360 does not apply to deliveries of plutonium extracted from non-US fuel – such as plutonium shipments between Britain and France – it is mandatory on any containers used by BNFL to return US-origin plutonium to Japan by air. Thus by the end of the 1980s, BNFL, like their French and Japanese counterparts Cogema and PNC, were attempting to design new plutonium containers to meet the US criteria.

Unlike magnox spent fuel flasks, plutonium containers are not tested in front of the nation's media. Rather the reverse: testing is highly confidential. Nevertheless a few details of earlier test programmes emerged in 1986 at the Dounreay Inquiry. In 1975, a drop test had been performed on an 0675 model plutonium package. The package comprised an inner canister (filled with a substitute material) packed inside a larger outer container. Dropped onto a concrete target from a helicopter at a height of 2,000 feet, the package hit the ground at just under 200 mph with an estimated force of 2,000 gs. The result would probably not have been given much publicity at the time: the force of the impact sheared off the bottom of the container and a small amount of powder leaked out of the inner can – contained, however, within the outer package.[34]

BNFL nevertheless kept quiet about an earlier test in 1973, whose results would have been far more interesting to Inquiry objectors. When a 0040 model plutonium package dropped from 2,000 feet, in BNFL's words (from a subsequent paper to an industry conference), it "disintegrated on impact, distributing its wreckage over a radius of 9m from the point of impact". The result of this embarrassing failure was a spillage of 4 per cent of the simulated contents of one of the four product cans in the package.[35]

In 1984, a more recent design of container, the 1676, was dropped from a height of 5,000 feet giving an impact velocity of 270-280 mph. This time, there was "less damage" to the container which survived without any internal leakage although it still did not meet the 288- mph impact criterion in Nureg-0360. As for fire tests, BNFL freely admitted at the Dounreay Inquiry that "the only test we have done is the standard IAEA fire test for 30 minutes . . .".[36] Responding to concerns raised at the Inquiry about a plane-load of plutonium crashing, BNFL's expert witness on the subject concluded, surprisingly perhaps: "It seems to me from the tests I have done (the 2,000 and 5,000 foot drops) . . . that the risks are very small, that it is likely there will be no leaks at all."[37]

When planning permission for a new reprocessing plant is at stake – not to mention hundreds of jobs and millions of pounds worth of contracts – one would not expect people paid to represent the applicant to express many doubts. But BNFL's public optimism has not been shared by everyone, even within the British nuclear industry. Six years before the Dounreay Inquiry, a UKAEA report for the European Commission reviewed the test criteria for plutonium containers – including the new US specifications – and came to more critical conclusions. BNFL might have been interested, for example, in the report's comments on fire testing: ". . . the IAEA fire specification is rather unsatisfactory. The temperature is low (by about 25%) for a hydrocarbon fuel fire."[38] "Such a fire would have a mean black body temperature of 1010°C and could last for a period of up to at least 2 hrs in a very severe accident."[39]

More significant, however, was the conclusion. Part of it is worth quoting at length:

> We feel that any final decision on appropriate test criteria for air-transportable plutonium containers should be based upon a risk analysis. Only by considering in detail the consequences of plutonium release can a rational comparison with other hazards be made. For it is clear that there will be a certain probability of plutonium release. The very severe nature of a minority of aircraft accidents indicates that complete protection (i.e. protection against the "maximum accident") is not possible. And in such severe accidents a breach of all containers in one payload is possible; very grave consequences could result.[40]

Clearly, within the privacy of an obscure technical document, UKAEA staff are prepared to concede a risk which BNFL's desire to reassure the public will not permit. Certainly the UKAEA report appears to contradict the optimism of BNFL whose expert witness at the Inquiry stated that: ". . . I have not seen any evidence which would change my view that air is a safer means of transporting plutonium than anything else."[41] Six years earlier, the UKAEA report had noted that ". . . potential accidents in the course of air transport are more severe than for other modes."[42] Nevertheless, although the UKAEA report recommended criteria more severe than the existing IAEA tests, they still fell short of the US specifications published in Nureg-0360. Whether the differences matter, of course, will only be put to the test when the first plane-load of plutonium crashes in a ball of fire.

The problems of designing new plutonium containers were further exacerbated by an amendment to Nureg-0360, introduced by Frank Murkowski, Republican Senator for Alaska. The "Murkowski amendment", approved in December 1987, stipulated that NRC approval of any new container for transporting plutonium must include two new tests: a drop test of a new container from a plane at maximum cruising altitude, and "an actual crash test of a cargo aircraft fully loaded with full-scale samples of such container loaded with test material". The necessity to crash an aircraft could be waived only if the NRC considered, after consultation with an independent scientific review panel, that the stresses on the container produced by other tests used in developing the container exceeded the stresses which would occur during "a worst case plutonium air shipment accident".

Murkowski's concern stemmed from proposals to fly plutonium from Britain to Japan via North America; flights on the long route to Japan have usually required a stopover, one option being Anchorage in Alaska. Unfortunately the USA's most northerly state has not always welcomed the nuclear industry. It was a 1971 underground nuclear test on Amchitka Island which helped give birth to Greenpeace. Local residents also remember the 1978 crash of a Soviet satellite, which contaminated parts of northern Canada with fragments of its nuclear power source: the ten-month operation to find Cosmos 954 and clean up its radioactive remains cost at least £6 million.[43] Perhaps recalling the saying "what goes up, must come down",

Alaska did not warm to its potential place in the plutonium economy. Reassurances about safe packaging were undermined by the memory of several dramatic accidents to nuclear weapons. Warheads accidentally dropped from aircraft over Thule in Greenland and at Palomares in Spain still managed, despite their sturdy construction, to spill plutonium and contaminate the surrounding environment. (See Chapter 9.)

In accordance with the Murkowski amendment, the NRC developed even more rigorous testing criteria, derived from a 1987 accident in the United States which had been identified as the "worst case" air crash on record. On 7 December of that year, a passenger killed the cockpit crew of Pacific Southwest Airlines flight 1771. When the British Aerospace-146 jet crashed out of control into a California hillside its estimated speed of impact was nearly 750 miles per hour. After analysis of the 1987 crash, it was proposed that the substitute for Murkowski's plane crash test could be a drop test that achieved a velocity of at least 630 mph – more than twice the speed of the impact test in the original Nureg-0360.

The difficulties of developing a suitable container for carrying plutonium made from US fuel and the problems of finding an acceptable route for its return to Japan have combined to make plutonium transport an international issue. Only two designs – the PAT-1 (Plutonium Air Transportable Package), weighing approximately 227 kgs, and the smaller PAT 2 – have been approved for use which meet even the original criteria of Nureg-0360. Both were developed in the United States by Sandia Laboratories to transport small quantities of plutonium; maximum capacities being 3.15 kgs and 15g respectively. By comparison, BNFL's plans – at least, prior to the Murkowski amendment – were to use the 1680 package with a capacity of 48 kgs. Compliance with Nureg-0360 would be achieved by incorporating a massive "overpack" to act as a shock absorber. The total weight of the package would be no less than 8 tonnes, half of which is accounted for by the overpack.

Whether such a large package for carrying substantial quantities of plutonium can actually be built to meet current US standards is debatable. Some experts have suggested that it is technically impossible: beyond a maximum size – a design threshold – the increasing weight of a package actually makes it more likely to collapse on impact. Moreover, even if a suitable container can be developed, the process of gaining NRC approval in accordance with the Murkowski amendment – the need to plan, carry out and evaluate the tests – will take an estimated four years to complete.

These factors, together with Japanese pressure for an early delivery of more plutonium (required no later than 1992 in order to reload Japan's Monju fast reactor) led to a United States compromise that must have relieved BNFL. As part of a US-Japan Agreement approved in October 1988, the Reagan administration agreed that if a suitable air transport container had not been approved by the NRC, plutonium made from US fuel could, subject to agreeing appropriate security arrangements, be returned to Japan by sea. This did not meet universal approval, even within the United States. Both the Department of Defense and the NRC argued against this backtracking on the grounds that security and safeguards were diminished. It nevertheless pleased BNFL – Japanese deliveries could go by sea. Deliveries to Europe – which would not necessarily be affected by US legislation – would, as originally intended, go by air.

The next shipment of plutonium to Japan is scheduled for autumn 1992. The plutonium will be transported on a modified freighter and escorted by an armed patrol boat purpose built for the Maritime Safety Agency (MSA), Japan's equivalent to a coast guard. There is no doubt that the years of low profile are over. Future deliveries of plutonium have become news even before they begin. Apart from attracting outright opposition, the nuclear industry will have to face probing questions from people and organisations who may not necessarily be anti-nuclear but feel they have a right to be concerned and informed. If large-scale transfers of plutonium are going to be driven and flown across Britain or shipped across the high seas, then elected representatives, local councils, and the emergency services – not to mention the general public – will inevitably want to know more about it. Whether the nuclear industry answers those questions is another matter.

Transport Hazards

Plutonium is one of the most carcinogenic substances known. The emission of alpha particles constitutes the main biological hazard. Although they travel only a few millimetres in air, their toxicity makes plutonium potentially fatal if it enters the body. These properties have contradictory implications for the design of plutonium transport containers. On one hand, alpha radiation needs no more than a minimal thickness of containment to prevent escape. On the other, plutonium's lethal toxicity demands a robust container with the highest standards of integrity; able to withstand extreme heat and impact in a severe accident and not release its contents.

The consequences of a release would depend upon the type of plutonium involved. Solid plutonium metal, which melts at 640°C, reacts in air acquiring a removable surface film of oxide. The metal is also pyrophoric and in the form of powder or fine chips, spontaneously catches fire if exposed to air to form a fine mist of plutonium oxide – a major hazard in the event of a release. Plutonium nitrate is a liquid and could be diluted or absorbed into whatever it came into contact with. In extreme circumstances, a leak could contaminate land or sea, or drain into sewers or public water supplies. Plutonium dioxide is usually transported as a powder. In this form, plutonium has a high melting point – 2,400°C – and is virtually insoluble in water. The main danger is inhalation, which depends on the size of the powder particles. These vary according to whether the material is granulated or milled. However, as a fine powder, plutonium dioxide could easily be inhaled or blown from the site of a spillage.

Transport arrangements for plutonium, a fissile material, must prevent unwanted criticalities. Transport packages make use of multiple containers, securely packed inside each other like a Russian doll. By surrounding the smallest innermost container with larger ones – usually made of steel packed with wood, cork or resin – the sheer bulk of the package provides the necessary distance to avoid criticality between adjacent containers. Such materials also provide insulation from external heat and shock absorbence against impact. These features are illustrated in current container designs, such as the 2816C "Safkeg" package, used by BNFL to transport plutonium dioxide.

A 2816C "Safkeg" package for transporting plutonium dioxide, on display at an industry conference. The major components of the flask (excluding potted plants) are the outer container on the right, the inner vessel on the left, and circular lid and lid plug in front.

Box 8.1: Plutonium containers

Plutonium dioxide is transported by BNFL in the 2816C "Safkeg" package. Designed by Croft Associates of Didcot and weighing about 116 kgs, the 2816C consists of two separate cylindrical containers, one inside the other; the outer container resembling an elongated beer keg – hence the catchy name. With an overall diameter of 425 mm and standing one metre high, the outer container is a double shell of stainless steel enclosing layers of cork insulation and expanded resin foam. The inner container, a deep stainless steel can accommodates two smaller cylindrical cans, one placed on top of the other. The smaller cans, of 1 mm thick stainless steel and sealed with welded lids, contain aluminium bottles with a capacity of 6 kgs of plutonium. The entire package is fitted with a stainless steel lid incorporating double "O' ring seals and secured by 13 steel bolts and a four-figure combination padlock. For bulk shipments, Safkegs are transported in metal stillages accommodating up to six Safkegs at a time.

To meet the requirements of Nureg 0360 – the result of reprocessing US-made fuel – BNFL have designed a new plutonium container for plutonium shipments to Japan. Designated the 1680, it offers a greater carrying capacity than the 2816C. The new container transports up to 48 kgs of plutonium and consists of three main components – small primary containers, secondary containers, and the outer packaging. Like the 2816C, the innermost compartment – the primary container – is a bottle with a screw-on lid which holds up to 6 kgs of plutonium dioxide powder. In the 1680 design, however, the bottle is made not of aluminium but 1 mm stainless steel and instead of a polythene bag, is sealed inside a 2 mm thick can with a welded-on lid, also made of stainless steel. This can in turn is sealed inside a second 1 mm thick stainless steel can, also with a welded-on lid – the welded lids of both steel cans are individually tested for leak tightness. The bottles of plutonium – each enclosed within two sealed cans – are then loaded into four stainless steel storage tubes within the 1680 package, each tube accommodating two canned bottles. The tubes are closed with a plug cap sealed with twin "O" rings. For insulation, shielding, and protection against impact, the four tubes are surrounded by timber with a minimum thickness of 400 mm. The entire assembly, with its cargo of eight bottles of plutonium, is enclosed by a mild steel shell and lid. For extra accident protection, an overpack shock absorber of wood encased in aluminium bolts onto the container which is transported in a horizontal position. The shock absorber alone weighs 5.4 tonnes and

Box 8.1 (continued)

together with the container, the complete package totals about 8 tonnes – all for the transport of 48 kgs of plutonium. The entire 1680 assembly – container plus shock absorber – is about 2 metres in diameter and 3 metres long.

New fuel for the PFR is transported in 3 tonne containers designated the "1356". This type of container holds a maximum of 4 subassemblies each of which is first loaded into an inner steel "tube" 3.2 mm thick. Four loaded tubes are then inserted into the outer box container made of 7 mm thick steel. The separation of the four subassemblies within the outer container is designed to prevent an unwanted criticality. The internal design of the container provides a further precaution against criticality: timber over 60 mm thick and 100 mm layers of polythene are built inside the outer container to protect and insulate the inner tubes. Overall the 1356 package measures 14 feet long by 3 feet square ands weigh about four tonnes when loaded.

The Seddon Atkinson vehicle used to transport Special Nuclear Material.

9   Nuclear warheads

On 10 January 1987, the chilling phrase "nuclear winter" almost acquired a new meaning. A convoy of military vehicles, reputedly carrying nuclear warheads, crashed on a public road in Wiltshire. Military vehicles are not an infrequent sight on our roads but this was no ordinary convoy. Sandwiched between dark green transits and motorcycle outriders were four "Mammoth Majors" – Leyland trucks specially designed for transporting nuclear warheads. Their cargo was assumed to be an unknown number of nuclear depth bombs en route to the Atomic Weapons Establishment at Burghfield. Fortunately, on this occasion no one was killed and, if the MoD is to be believed, no radioactive material released.

That winter had been particularly bad. January was a month of snow and ice. News bulletins urged motorists not to travel "unless your journey is really necessary". Such reservations have no place within the Ministry of Defence. The ability to deter the Red Army (or whoever the latest enemy is) from sweeping across Europe – even in a blizzard – requires eternal vigilance. Britain's nuclear deterrent cannot stay at home because the weather's bad; warheads must be at the ready and, as necessary, on the move.

At 3.15 on this particular afternoon, an indeterminate number of them were parked on the outskirts of Southampton; the convoy had stopped on a slip road at the junction of the M27 and M271. Heading north west, it had travelled from Portsmouth whose naval base is regularly used by ships which can carry nuclear weapons. The convoy headed for the village of West Dean on the Wiltshire/Hampshire border and the neighbouring Royal Naval Armaments Depot at Dean Hill. On the face of it, Dean Hill does not look much of a base. A notice at a side entrance instructs personnel to "SEARCH YOURSELF BEFORE ENTRY" – hardly a sign of top secrecy. On the other hand, none of the base's buildings are marked on the local ordnance survey map, nor is its private railway system which passengers on the line between Salisbury and Romsey can see branching off near East Dean. Officially, convoys of nuclear warheads don't exist either, so perhaps it's appropriate that Dean Hill has provided a convenient location for an overnight stop on the journey to Burghfield. As the third of the Mammoth Majors overtook a parked car on the icy road to West Dean, it slid out of control and skidded off the road. The vehicle and its nuclear cargo crashed through a nearside hedge, plunged down a shallow bank, and landed on its side in a field of snow. The following warhead carrier also skidded into the hedge but stopped short of the bank with one wheel off the road.

This certainly wasn't the first accident affecting a Mammoth Major, nor the first time that the Ministry of Defence refused – as always – to confirm or deny the presence of nuclear weapons. Nevertheless, the resolute response at West Dean was hardly compatible with a crash of empty vehicles: if nuclear warheads had not been present, then the Ministry of Defence turned the incident into a remarkably realistic rehearsal.

The West Dean accident was something of a turning point. For the first time in Britain, the transport of nuclear warheads made the national news headlines. The incident highlighted not only the physical reality of the nuclear deterrent but also the dangers of nuclear weapons in peacetime. Nuclear warheads are in effect transport containers enclosing a mixture of conventional high explosives and radioactive materials: fissile ingredients (plutonium and highly enriched uranium), tritium (a radioactive isotope of hydrogen), and depleted uranium. All of these substances are, to a greater or lesser extent, potentially hazardous – especially plutonium – and their release or exposure to the environment could be a public health disaster.

British nuclear warheads are deployed by ships, submarines and aircraft (although they are not carried on combat aircraft or ships in peacetime.) Warheads must therefore be moved from AWE Burghfield, where they are assembled, to the locations from where they are deployed; by road to bases within Britain, and by RAF transport aircraft to bases overseas. The frequency of warhead movements is a product of Burghfield's output of new models, the rate at which the Ministry of Defence retires old ones, and the need to service operational warheads during their lifetime – warheads are routinely withdrawn from service and returned to Burghfield for a "checkup". In addition, prototypes and service models are periodically transported to the United States for test explosions.

Britain's nuclear deterrent

At the end of the 1980s, Britain's nuclear arsenal consisted of three principal types of warhead: the Chevaline A3-TK warheads carried by Polaris submarines; the WE-177 tactical nuclear warheads, a free fall bomb carried by the RAF; and the Royal Navy's nuclear depth bomb, a variation of the WE177. In addition to these British-built warheads,

British forces have also had access to US nuclear warheads. However, US weapons available to British forces have been deployed in Europe with one exception: stocks of the US B57 nuclear depth bomb are held in Britain for anti-submarine use by RAF Nimrod aircraft.

Warhead movements invariably increase as new models are introduced and old ones withdrawn. In July 1980, the government announced its decision to replace Polaris with four new submarines carrying US Trident missiles. The first, HMS Vanguard, is due to enter service in 1994. Each boat will be capable of carrying 16 US Trident missiles each of which can be armed with up to 14 independently-targetable nuclear warheads – a potential capacity of 896 warheads. In fact, the government has stated that the submarines will carry a maximum total of 128 warheads per boat[1] – i.e. an average of up to only 8 warheads per missile. Despite this apparent restraint, Trident is not merely an updated replacement of Polaris but a considerable expansion of nuclear fire-power, as the government itself has acknowledged: "this (Trident) represents an increase of up to 2.5 times the payload of the Polaris boats, when they entered service in the 1960s".[2] Production of the warheads began at Aldermaston in January 1988.[3] The other development is a replacement for the WE177 which is nearing the end of its life. This is likely to be an air-to-surface nuclear missile developed in collaboration with France or the United States.

Despite government secrecy about the transport of nuclear warheads, there is no secrecy about the delivery systems – the planes, ships and helicopters which can carry them nor where they are based. It is therefore easy to identify the bases which handle nuclear warheads, which, by extrapolation, gives a broad outline of transport routes to and from Burghfield. These can be summarised as follows:

1 Burghfield to the Royal Naval Armament Depot (RNAD), Coulport, Loch Long, Strathclyde. Coulport is where Chevaline warheads are stored for deployment on Polaris submarines. Coulport will also store Trident warheads.

2 Burghfield to RAF Honnington, East Anglia. Honnington is the base where WE-177 warheads are stored for Tornado bombers. The most frequented route, with one or two convoys per week transporting WE177 free-fall warheads to and from the Tornado base in East Anglia. Honington is the base from where nuclear warheads are flown to and from RAF bases in Germany.

3 Burghfield to RNAD Frater, Portsmouth. Portsmouth is the home port for the three Anti-Submarine Warfare carriers, which carry Sea Harrier aircraft and helicopters capable of deploying WE-177 warheads and British nuclear depth bombs.

4 Burghfield to RNAD Bull Point, Plymouth. A convenient storage

depot for nearby Devonport, home port for nuclear capable frigates and destroyers.

Not all convoys are to or from ROF Burghfield; warheads also travel between nuclear bases, for example, between Honnington and Portsmouth. (One explanation for this practice is that the WE177 is redeployed as its tritium inventory decays. This would reduce its potential yield making it more suitable for use by the Navy as a depth bomb.)

All convoy routes are varied for obvious security reasons. The avoidance of regular habits is basic security practice and one reason why road transport is preferred to rail – "the fixed nature of a rail route might be undesirable".[4] Convoys nevertheless stick to a limited number of main roads and motorways where ever possible. However, while the use of more than one route may well be a desirable security measure, it increases the number of people and places potentially at risk in the event of an accident endangering a warhead. One has only to plot convoy routes on a map to appreciate how many towns and cities lie close to warhead routes. An accident to a nuclear warhead could happen almost anywhere. . .

Motorway Madness – the arms race on the move

The "Mammoth Majors" were introduced at the beginning of the 1980s to provide a delivery service for the warheads of the RAF and the Navy. Previously, for example, Polaris warheads had been shipped by the Navy's Royal Fleet Auxilaries from Coulport to Plymouth and Portsmouth from where they were delivered to Burghfield by road. The practice of shipping nuclear warheads through the Irish Sea at the time of an IRA revival was ostensibly one reason why transport arrangements were changed.

The Mammoth Majors themselves are not unduly distinctive. More conspicuous as a clue to their cargo is the formation of the whole convoy. RAF police motorcycle outriders and vans carrying armed guards flank the warhead carriers in front and behind. The convoy also includes two specialised support vehicles: a fire tender, which follows the last Mammoth Major, and the Convoy Support Vehicle Coach – a mobile control centre bristling with aerials. Latterly, convoys have also included a breakdown truck for towing. In the middle are up to five Mammoth Majors, the last of which is always empty in case it is needed as a spare.

The introduction of the Mammoth Majors was accompanied by a new unit – the RAF Special Safety Organisation – set up to control the convoys. Based at RAF Locking near Weston-super-Mare, the SSO brought together officers of the RAF Regiment, the RAF Police and the Commachio Company of the Royal Marines. The Commachios – also stationed at

the Faslane Polaris base and at the Navy's nuclear armament depots –
are a special security force set up in 1980. Along with members of the
RAF Regiment, they provide armed guards to protect the convoy. The SSO
itself is controlled by the MoD's Nuclear Accident Response Organisation
(NARO) which in the event of a major incident, can call on between 20
and 30 specialist military units around the country, trained to respond to
emergencies.[5]

A convoy of Mammoth Majors transports Polaris warheads from Coulport, Scotland
to AWE Burghfield, Location: M6 near Preston

Apart from West Dean, Mammoth Majors have been involved in at least
two other accidents. On 20 June, 1985, two Mammoth Majors collided with
each other in Helensburgh, near Glasgow. Sandwiched between hill and
loch, Helensburgh is just down the road from the Faslane submarine base
and its nearby weapon store at Coulport. There had been signs of trouble
even before the accident. Eye witnesses also reported smoke coming from
the rear axle of one of the warhead carriers. As it drove through the town
centre, the recalcitrant vehicle smashed into the back of the preceeding
Mammoth Major. The convoy came to a halt. RAF motorcyclists with the
convoy bawled at motorists and pedestrians and sealed off one side of the
crowded street, "They looked in a bit of a panic," said one onlooker.[6] The
accident had happened in Helensburgh's busy shopping area, just down the

road from a Fine Fare supermarket. After years of secrecy, the Ministry of
Defence had inadvertently provided the press with the perfect photo call.

A Mammoth Major on the M6 near Carlisle

Despite the apparent drama, the only visible signs of damage were broken
windscreens on one of the Mammoth Majors. After 20 minutes, the vehicle
was towed back to Faslane for repairs. More damaged by the accident was
the MoD's credibility. Despite the tacit assumption by everyone else that
the convoy carried nuclear warheads, the Ministry of Defence continued to
neither confirm nor deny their presence, a policy that has been increasingly
unconvincing. Parliamentary questions led to exasperated replies: "I should
have thought it was obvious that if the Ministry of Defence gave details of
accidents involving nuclear weapons it would be disclosing the methods of
transportation".[7]

Another accident made the news in 1988. On 17 September, a convoy
transporting nuclear depth bombs from Plymouth to Burghfield collided
with a car on the A303 Ilminster by-pass in Somerset. When a coach
tried to overtake the convoy, the car, travelling in the opposite direction,
swerved across the wrong side of the road in front of the oncoming vehicles
and collided head on with a Mammoth Major. The car broke in half and the
driver, medical student Jonathon Chubb, died soon after from his injuries.

During the second half of 1992 the Mammoth Majors are due to be
replaced by what are officially described as "Truck Cargo Heavy Duty

Mk 2" vehicles, which will transport all current and future nuclear warheads, including those for Trident missiles. Eight of these specially designed articulated vehicles (plus one additional tractor unit), have been ordered by the MoD from Root Brown Vickers.[8]

Servicing nuclear warheads

The movement of nuclear warheads is not limited to the introduction of new models and the withdrawal of old ones. At periodic intervals during their life, operational warheads must be returned to Burghfield for a checkup, as the government has acknowledged: "British nuclear warheads are designed to be safe and serviceable. Their planned withdrawal from service for inspection and periodic maintenance ensures that this aim is met".[9] The need to remain "serviceable" accounts for many of the movements between the bases and depots which handle nuclear weapons and AWE Burghfield, where inspections are carried out. Such inspections are necessary partly because the radioactive materials in warheads decay.

Tritium, in particular, presents problems. It provides part of the thermo-nuclear fuel for "boosting" a warhead's yield. It also has a relatively short half life of 12.3 years, decreasing in quantity by 5.5 per cent each year. With the WE177, the need to replenish tritium components requires returning the entire nuclear warhead to Burghfield. More recent designs such as Chevaline reportedly have removable tritium components which can simply be taken out and replaced.[10] This innovation reduces the need to transport the complete warhead around the country merely to renew one ingredient. Tritium is returned to Chapelcross by road from either Coulport – where Chevaline warheads are stored – or Burghfield. The transport of tritium is assumed to be one of the functions of the SNM convoy described in the previous chapter.

United States nuclear weapons

Most of the nuclear warheads deployed in Britain over the years have not been British but American. In recent years the most notorious were those that came with the Cruise missiles at Greenham Common and Molesworth. However, other US bases have been operational for far longer and held larger stocks of warheads. Deployment has been as follows:

USAF Lakenheath: 300 warheads for use by F-111 bombers
USAF Upper Heyford: 300 warheads for use by F-111 bombers
Holy Loch, Scotland: base for 10 US Navy Poseidon submarines capable

of carrying a total of up to 2,240 warheads. With 3 submarines usually in port at any one time and assuming an average loading of 10 warheads per missile (with 16 missiles per submarine), the Clyde base typically contains 480 US warheads.

However, at the time of writing, major changes are afoot. Holy Loch is to be wound down and by the end of 1994, the F-111s at Lakenheath and Upper Heyford will also have gone. Nevertheless, while Upper Heyford will revert to standby, Lakenheath will continue as a nuclear base with the arrival of the F-15E. This new tactical fighter bomber will inherit the B61 gravity bombs currently available for the F-111. Thus while the aircraft change over, Lakenheath's bombs will remain. In the longer run, the F-15E may be allocated the Tactical Air-to-Surface Missile (TASM), one of the fruits of NATO's nuclear modernisation programme.

Transport arrangements for US warheads differ markedly from British bombs. Warheads are stockpiled at the US bases which deploy them and transported by air and sea between Britain and the United States, where they are assembled, dismantled and serviced. Air deliveries are made via the US Air Force base at Mildenhall in Suffolk, from where they are flown on to Lakenheath and Upper Heyford. Mildenhall is also a stopover for aircraft collecting and delivering B57 warheads for the RAF at St. Mawgan (where the B57 is deployed) and for US bases in mainland Europe. Holy Loch receives regular visits from two US supply ships, the *Marshfield* and the *Vega*, which transport Poseidon warheads across the Atlantic. These deliveries will presumably cease when the base closes at the end of 1992.

These arrangements mean that US warheads should not need to move around on British roads. However, there may be exceptions. Mammoth Majors have in the past been seen, for example, in the vacinity of RAF St. Mawgan, the Welsh Nimrod base which stores US depth bombs for British use.

While the absence of US nuclear warheads on British roads might seem like a safety advantage, transporting them by air is far more dangerous – US warheads flying through British airspace are at risk from accidents in the air, and on landing and takeoff.

Accident potential

It's hard to imagine a more dangerous environment for radioactive material: not only do nuclear warheads contain some of the most toxic radioactive materials known, but they are encapsulated cheek by jowl with high explosives. In any other situation, strenuous efforts would be made to keep such hazards apart. And yet moving around at sea, in the air and on our

roads are weapons which integrate within a confined space both radioactive
materials and high explosives (HE).

As with other radioactive materials in transit, the contents of a nuclear
warhead are vulnerable to fire and impact. Fire has generally been regarded
as the more serious danger. While a severe impact is only likely in a
major accident, a serious fire can potentially develop in a wider range of
circumstances – especially during road transport – and from the most trivial
of beginnings. Any fire that threatened the radioactive components could
also threaten the conventional explosives: it takes no breach of military
secrecy to appreciate that non-nuclear explosives when heated have a
tendency to explode – thus blasting dangerous radioactive substances into
the surrounding environment.

Various accidents over the years have shown that an impact, if severe
enough, can detonate the conventional explosives in a warhead and blow it
apart. Warheads can also be ruptured by impact alone. Several well-known
accidents have shown that either can result in the release of radioactive
material. On 23 January 1961, a wing fell off a US B52 bomber flying over
North Carolina near Goldsboro. On board were two 24-megaton warheads
which crashed to the ground with the plane. One of the bombs split open
on impact, releasing uranium and contaminating farmland to a depth of 50
feet. Five years later, the *Washington Post* reported that part of the bomb
had still not been recovered – it remained buried somewhere in the North
Carolina soil.[11] In 1966, a US B52 and a KC-135 refuelling tanker collided
in mid-air over Palomares in Spain. When two of the bomber's four nuclear
warheads hit the ground, their conventional explosives detonated on impact
scattering plutonium across a wide area. (A third warhead fell in the sea
while the fourth hit the ground without exploding.) Two years later another
B-52 carrying four nuclear warheads crashed near Thule in Greenland. The
conventional explosives in all four weapons detonated, strewing plutonium
over the surrounding area.

An earlier accident in England highlighted the dangers of fire. On 27
July, 1956, a US B47 bomber crash landed and burst into flames at the
Lakenheath base in East Anglia. While the plane itself was not carrying
nuclear weapons, flames from its blazing fuel tanks engulfed a nearby
bunker containing nuclear warheads. The crash was briefly reported at the
time but it wasn't until November 1979 that a local US newspaper published
a fuller account of the story. According to the *Omaha World-Herald*, the
heat severely damaged the bombs and came close to detonating their TNT
explosives. Although the fissile cores had been removed from the warheads,
they were stored in the same beleaguered bunker. The apparent likelihood of
a TNT explosion scattering radioactive material over the surrounding area
led to a mass exodus from the base of panic-stricken personnel.[12]

Seventeen months after the crash, Prime Minister Harold Macmillan assured parliament that: ". . . there is no danger of the hydrogen bomb exploding in the event of a crash of the aeroplane."[13] In the light of the belated revelations about Lakenheath, such reassurances at best look disingenuous. It is now quite clear that the dangers of a fire causing an accidental *conventional* explosion in a nuclear warhead were recognised at that time, although they weren't admitted in public. In the early years of the bomb programme, the Ministry of Defence carried out considerable research into the vulnerability of warheads to accidents. Some of this formed part of the "Maralinga Experimental Programme" carried out in Australia in the 1950s and early 60s. In addition to several full-scale nuclear explosions, Britain conducted a series of "minor trials" including a programme of experiments code-named "Vixens" carried out at Maralinga between June 1959 and May 1963.

The Vixen trials were designed to assess the effects of accidents on nuclear warheads, reportedly using 60 kgs of plutonium flown out from Britain.[14] The first phase of the programme, Vixen A, examined how plutonium and other toxic materials would disperse after a fire or explosion. These trials included experiments where plutonium was deliberately burnt in petrol. In a second series of trials, designated Vixen B, "live" nuclear warheads – complete weapon assemblies containing plutonium – were destroyed by detonating the conventional high explosives to deliberately release fissile material. While the safety of British nuclear weapons may well have benefitted from these dubious experiments, the fact that it was considered necessary to conduct them at all – and in the wilds of remotest Australia – shows that the accident potential of nuclear weapons has at least been recognised by those who actually build them. It is also now recognised by the government. In 1990, for example, Home Office advice to fire service personnel noted that:

> If a release of radioactivity did happen, this could be as a result of one of the following:
> a. Extremely severe impact damage, resulting in damage to the weapon casing and its protective container, and leading in turn to exposure of the radioactive material which could then spread elsewhere.
> b. A fire leading to fire or explosion of the conventional explosive within the weapon and so in turn to a scattering of the radioactive material. The spread would be confined to the smoke plume in any fire and the explosive fragment area.

The Home Office also drew attention to another potential danger in the event of an accident. Radios, it cautioned, should not be used within 10 metres of any weapons because "there is a low possibility that a radio transmission could cause detonation of the HE present".

The worst conceivable accident, however, is an unintended nuclear detonation. Officially this has always been deemed impossible: warheads have to be armed before a nuclear explosion can occur. During transportation, for example, warheads are normally maintained in an unarmed state achieved by disconnecting the electronic firing components. Arming a warhead is a complex procedure requiring the coordinated actions of several people; various mechanical and electronic safety measures would have to be overridden before a chain reaction could be initiated. It should therefore not be possible for an accident to unintentionally arm a nuclear bomb. Moreover, warheads are designed to ensure that if even the conventional explosives detonate accidentally, they blow the weapon apart rather than start a nuclear blast (known as "one-point" safety). Thus the British government has confidentally asserted that: ". . . the inadvertent or accidental arming of (nuclear) weapons is not possible."[15]

Yet occasionally the impossible is qualified, allowing a hint of doubt to emerge. In 1987, for example, a US report noted that "extensive safety precautions have made the probability of an inadvertent nuclear weapon detonation virtually non-existent".[16] Note the qualifying word "virtually": the difference between "non-existent" and "virtually non-existent" need not necessarily be semantic. The bombs which crashed near Goldsboro had been in a armed state during flight and some twenty years after the event, former US Defence Secretary Robert McNamara revealed that the crash had triggered five of the six electronic arming devices in one of the bombs: the conventional explosives had been just one trigger away from an accidental detonation. While this would not necessarily have triggered a full-blown nuclear blast, it might have resulted in a "partial fizzle" and would certainly have dispersed radioactive material into the environment.[17]

It has generally been assumed that safety precautions have improved since those early days of the arms race; for example, US bombers no longer carry nuclear warheads in an armed state of alert. Yet a 1990 study by a panel on nuclear weapons safety convened by the United States House of Representatives Armed Services Committee included one of the most startling qualifications to date. The Drell report, as it is known, noted how advances in computer modelling techniques had enabled more realistic forecasts to be made of how a nuclear detonation might occur. It concluded: "A major consequence of these results is a realization that unintended nuclear detonations present a greater risk than previously estimated (and believed) for some of the warheads in the stockpile."[18]

The report noted several ways in which the possibility of an unintended detonation caused by an accident can be reduced. Warheads can incorporate "insensitive high explosives" (IHE) which are far less likely to explode on impact or in a fire compared to conventional high explosives (HE). Other safety measures include the Enhanced Nuclear Detonation Safety (ENDS) system, where the electronic components that arm a warhead are physically isolated from the rest of the weapon, and the fire-resistant pit (FRP), an innovation which further reduces the likelihood of plutonium dispersal in the event of a fire. The report recommended that all US nuclear weapons be equipped with ENDS and the development of warhead designs where the plutonium capsule is physically separated from the IHE prior to arming the weapon.

Transport risks were also considered. Noting that the US Department of Energy transport by air only warheads with IHE (unlike the US Department of Defence), the panel recommended that all nuclear bombs loaded onto aircraft should incorporate IHE and fire-resistant pits: "These are the two most critical safety features currently available for avoiding plutonium dispersal in the event of aircraft fires or crashes."

Doubts were cast by the panel on the safety of two US weapons systems in particular: the W69 warhead of the SRAM-A missile – which had been the subject of previous safety concerns and had been "taken off the alert" by the US Defence Secretary in June 1990 – and the W88 warhead being built for the US Trident D5 missile. At the time of writing, neither are currently based in Britain, but the report's criticism of the US Trident warhead – it has neither IHE nor a fire-resistent pit – begs obvious questions about the safety of its British equivalent. Does the British Trident warhead have insensitive high explosives? If not, how vulnerable are its conventional high explosives to accidental detonation? Does it have a fire-resistant pit? Such questions might also be asked about Britain's existing nuclear warheads, the Chevaline A3-TK and in particular the WE177 – the warhead involved in the West Dean and Ilminster accidents. The WE177 design is based upon the US B57, an old warhead identified by the US Department of Energy as lacking modern safety features and deployed in Britain for use by RAF Nimrods. The British government's response has been predictable. Asked if British warheads use insensitive high explosives, the Defence Secretary stated: "It would not be in the national interest to reveal such details of nuclear weapon design".[19] In the absence of such details, the transport of nuclear warheads leaves at best many unanswered questions; at worst, a potential disaster.

Accident consequences

With nuclear warheads moving around Britain by road and air, the possibility of fissile material being dispersed into the environment raises legitimate questions about public health and safety. Yet while there can be no guarantee that such an accident will not occur the consequences are consistently downplayed.

In 1988, for example, the Defence Secretary was asked what assessment he had made "of the potential consequences for the local population of an accident to a nuclear weapons convoy". Armed Forces Minister, Archie Hamilton, replied: "In view of the extensive safety arrangements involved in the design, construction, containerisation and transport of nuclear weapons, our assessment is that it is inconceivable that the impact which could occur in the worst type of road accident would be sufficient to lead to any release of radioactive contamination into the environment."[20]

Undoubtedly the threat to a warhead from any impact is far less in a road accident than it would be if a bomb fell out of the sky. Nevertheless, Hamilton's reply studiously avoided any reference to the hazards of fire which had been highlighted in two convoy crashes earlier that same year. When a Mammoth Major hit Jonathon Chubb's car near Ilminster, the impact spilled petrol from the car onto the road although luckily it did not ignite. Fire was also a potential problem at West Dean. The fire tender at the rear of the convoy had been prevented from reaching the overturned Mammoth Major by another warhead carrier in between which had skidded and blocked the road. Had a fire broken out, it could have burned out of control. Clearly there is no limit to the number of scenarios in which road accident fires can develop.

Both plutonium, and to a lesser extent highly enriched uranium, will burn in the presence of air (to form plutonium and uranium oxides) and change their physical states. Plutonium can form an aerosol, a cloud of minute particles which could drift downwind from an accident. Plutonium particles could be inhaled or ingested from direct contact with either a plume of dispersing material or contaminated ground. The consequences for public health would depend upon the nature of the accident. In the absence of any published information from the British government, the Drell report outlined the relative dangers:

> ... there is an important difference between dispersing plutonium via a fire, or deflagration, and via an explosive detonation. In the latter case the plutonium is raised to a higher temperature and is aerosolized into smaller, micron-sized particulates which can be inhaled and present a much greater health hazard after becoming lodged in the lung cavity. In the former case fewer of the particulates are small and readily inhaled; the larger particulates, although not readily inhaled, can be ingested, generally passing through the human gastrointestinal system rapidly and causing much less damage. As a result, there is a difference by a factor of a hundred or more in the areas in which plutonium creates a health hazard to humans in the two cases.[21]

The report calculated that the detonation of the high explosive in a typical nuclear warhead would contaminate an area of approximately 100 square kilometres downwind of a release. In an earlier US government report computer models were used to predict the dispersal of a notional plume of plutonium. While the most severe contamination might be limited to the area immediately adjacent to the accident, people up to 35 kms away could be exposed to a potentially harmful dose.[22] Another assessment has calculated that under the most unfavourable accident conditions, a "significant radiological impact" could extend up to 200 kms from an accident site.[23] If that sounds too hypothetical, one has only to recall the now famous reaction of the firefighters at the Lakenheath crash: the US personnel were ordered to turn their fire hoses on the burning bomb store at the expense of the aircraft's crew, who perished in the blazing plane.

In Britain, however, the hazards of plutonium are consistently downplayed. Back in 1957 for example, Prime Minister Harold Macmillan, responding to concern about warhead accidents, summed up the dangers of burning plutonium and uranium: ". . . hazards arise really because these are materials which . . . emit alpha particles . . . but the range of alpha particles in the air is very short, in fact less than one millimetre; It can therefore be seen that any hazard from this form of radiation is very slight. The possibility of ingestion and of external hazards is really negligible."[24]

What Macmillan failed to point out were the dangers of inhalation and the *internal* hazards of alpha particles. A similar omission was made more recently in relation to training exercises for warhead accidents. Some have involved the dispersal of radioactive isotopes to simulate the release of materials from a warhead. Exercises practising, for example, decontamination have been carried out at military bases around the country. (One was at RAF Swynnerton, Staffordshire, where radioactive material was used to contaminate an area around a simulated accident site to a radius of half a mile.[25] Coincidentally – or perhaps not – Swynnerton is one of the overnight stopovers used by warhead convoys travelling between Burghfield and Coulport.)

News reports of these exercises, and the use of radioactive contaminants, provoked questions in parliament, especially from MPs in the constituencies concerned who predictably knew nothing about them. By contrast, many were familiar with the Australian government's Royal Commission of Inquiry into British nuclear tests in that country, which had concluded its hearings two years before. Those investigations revealed, *inter alia*, that radioactive materials had been used in accident tests (the Vixen trials) and left a legacy of contamination.

The British exercises had used dispersed contaminants applied in heavily diluted liquid, powder or pellet form; radium-223 sulphate was a typical example.[26] Its half life of 11.4 days was supposedly the longest of any of the substances used.[27] Reassurances by the Defence Secretary were in the tradition of Harold Macmillan: "The damaging effect of the ingestion of radioactive material by the Service men or civilians involved who were not wearing protective clothing would be broadly equivalent to smoking one cigarette."[28] Notwithstanding government health warnings about cigarette smoking, radium-223 is an alpha emitter and like other alpha emitters the principal hazard comes not from ingestion but from inhalation.

Attempts to downplay radiation hazards – even from a training exercise – put a question mark over how the consequences of a major accident would be dealt with. If a release of radioactivity led to widespread contamination, would the Ministry of Defence clear it all up? Past experience overseas does not encourage optimism. For example, the cleanup operation after the B52 crash at Palomares involved removing contaminated top soil – but only over a certain level of radioactivity. This led to a conflict between the Spanish and US authorities. The Spanish government pressed for the removal of all soil that gave a geiger counter reading of over 7,000 counts; soil over 700 counts should be ploughed under to reduce the danger. The US authorities disagreed – and got their way: only soil over 50,000 counts was removed (1,700 tonnes transported to the US); soil between 7,000 and 50,000 was ploughed under, while a count below 7,000 was irrigated.[29] Within three years of the crash, illnesses among local people and livestock had already been reported.[30]

The difficulties of removing every last particle of plutonium – not to mention the cost of decontamination – suggest that an accident in Britain would not be totally cleaned up. But that of course is speculation: it goes without saying that contingency plans are top secret. For those outside the MoD who might be involved in dealing with the consequences, particularly the emergency services and local authorities, such plans may as well not exist. The MoD's policy of revealing as little as possible inevitably raises more questions than it answers. Perhaps the most important is why out of 160 or so countries who are members of the United Nations, a mere half a dozen or so feel unable to defend themselves without nuclear weapons?

10 Radioisotopes and other radioactive materials

It is something of a paradox that the materials whose transport has generated most controversy account for only a small proportion of the total number of movements of radioactive materials. These include radioisotopes of one sort or another, which account for by far the largest number of journeys made by radioactive materials. They are used widely in medicine, industry and research and even in the home. On a personal level, they move around on the wrists of millions of people whose watches have luminous dials – the radioisotope tritium provides the glow in luminous paint. At the other extreme, radioisotopes (plutonium-238) have travelled to the edge of the solar system supplying power for spacecraft like *Voyager-2*.

The term *isotope* was coined in 1913 by Sir Frederick Soddy, a lecturer in the chemistry department at Glasgow University. Soddy observed that while atoms of the same element all had the same number of protons in their nucleus, some had a different number of neutrons. Thus the nucleus of uranium-235 has 92 protons and 143 neutrons while uranium-238 has 92 protons but 146 neutrons. At the other end of the spectrum, the nucleus of hydrogen, the simplest of all elements, comprises just a single proton. However, there are also two other kinds of hydrogen: deuterium, whose nucleus has a proton and one neutron, and tritium, with one proton and two neutrons – deuterium and tritium are the other two isotopes of hydrogen. Isotopes can be radioactive or stable. Tritium, for example, is radioactive while deuterium is stable.

Of the 92 naturally-occurring elements most are made up of several different isotopes; more have been created artificially. Most of those which have been exploited for commercial use are produced from non-radioactive materials either by irradiating them in a nuclear reactor or by bombarding them with sub-atomic particles in a cyclotron. A brief summary of the different uses illustrates some of the radioisotopes which move around and who the customers are.

Research and medicine

Radioisotopes have been increasingly exploited in what are broadly termed the life sciences. Research in biomedicine, biochemistry and molecular biology – in universities, research institutions and drug companies – frequently utilises organic compounds incorporating a minute trace of a radioisotope; typically carbon-14 or tritium. The behaviour of these "labelled" compounds can be tracked by radio-tracing.

In medicine, the best known use of radiation is probably radiotherapy. Various forms of cancer are treated by irradiating malignant cells with gamma radiation from a radioactive source. In *teletherapy*, gamma radiation, usually from cobalt-60, is beamed onto a localised area of a patient's body. Another form of treatment – *brachytherapy* – utilises substances like caesium-137 which are implanted into the body to irradiate tumours internally.

Radioisotopes are also used diagnostically. The behaviour of a radioactive substance within the body can be monitored externally to show how the body is working. Such applications utilise short-lived radioisotopes like rubidium-81 which emits krypton-81m as it decays. Krypton-81m is a gas (with a half-life of just 13 seconds) used to diagnose a patient's breathing problems: after inhaling the gas, the patient's chest is photographed by a "gamma camera" to give doctors a picture of how the lungs are ventilating. The most widely used radioisotope for diagnostic purposes is technetium-99m. This is a decay product of molybdenum-99 supplied to hospitals in kits known as "technetium generators"; the technetium is eluted from the kit as needed.

Industry

Radioisotopes for industrial use fall into two broad categories: sealed and unsealed sources. Sealed sources comprise an isotope emitting gamma or beta radiation whose penetrating properties provide a means of making measurements without physically interfering with the object or material being measured. Applications include thickness, level and density gauges. Sealed sources containing beta emitters like strontium-90 and krypton-85 measure the thickness of materials like paper and plastics. Gauges incorporating gamma emitters such as caesium-137 measure the level of liquids in containers or materials in storage tanks. Well-logging provides another use of caesium-137. This involves measuring rock density in oil and gas wells during drilling and production to obtain geological information about the surrounding strata. Sealed sources emitting gamma radiation are widely used in radiography. Typically employing iridium-192 sources, radiography is analagous to an X-ray and enables industrial structures like pipelines and concrete bridges to be checked on site for quality control.

Unsealed sources comprise radioisotopes which are released into industrial processes for use as "tracers". By tracking the spread or progress of the radioactivity, the flow of liquids can be measured and leaks detected in pipework, for example, in petrochemical plants.

Gamma irradiation units

One of the more controversial uses of radioactivity has been to irradiate food. Gamma irradiation kills bacteria and can be used for disinfestation, to inhibit sprouting, delay ripening and prolong shelf life. It has also been widely criticised and until 1990 the irradiation of food for human consumption was prohibited in Britain. In other countries, however, cobalt-60, and to a lesser extent caesium-137, are used to irradiate a surprising variety of food products, primarily for insect or microbial disinfestation.[1] The ability to kill micro-organisms has other applications and irradiation plants have been used to sterilise medical equipment like hypodermic syringes and needles and such diverse items as plastic bottles, dog chews and corks. Materials like wood and plastic can be irradiated to change their physical properties: irradiation can alter molecular weight, affect melting properties and improve material strength.

Four commercial gamma irradiation plants currently exist in Britain, operated by Isotron plc. Two are located next to each other in Swindon, the others are in Bradford and Slough. Isotron's facilities use massive amounts of cobalt-60, produced by irradiating non-active cobalt-59 in nuclear reactors. The major world producer is Ontario Hydro in Canada which manufactures cobalt-60 in its Pickering and Bruce nuclear power stations. The irradiation of food and industrial and medical products is therefore a by-product of the nuclear fuel cycle.

Military uses

Although radioisotopes are mostly seen as a beneficial use of the atom – even by those who are hostile to the rest of the nuclear industry – they do not always meet universal approval. Some play a small but vital part in Britain's nuclear deterrent. Amersham International, for example, have supplied Rolls-Royce and Associates with neutron sources for nuclear submarine reactors.[2] (Neutron sources are radioactive "pilot lights" which start up the fission process inside a reactor.)

Other radioisotopes are used by the Ministry of Defence for "health and safety" purposes. Cobalt-60 is utilised in instruments for measuring radiation, for example in the naval nuclear dockyards. Cobalt-60 has another military application that should come in handy in a nuclear war: it is used in instruments employed by the Royal Observer Corps to callibrate equipment for measuring fallout.

Radioisotopes have also been used in MoD training exercises for nuclear weapon accidents. Radioactive substances with relatively short half lives have been released in MoD training areas to simulate the dispersal of materials like plutonium. Over the years, the MoD has used compounds of radium, technetium-99m and yttrium-90 oxide. Callibration of the MoD's equipment for measuring the resulting contamination requires another selection of radioisotopes, typically caesium-137, chlorine-36, cobalt-60, iodine-131, plutonium and strontium-90.

Radioactive isotopes are also likely to be used by US forces in Britain, accounting for further movements of radioactive materials. (The US Army in Germany, for example, receives a variety of radioactive materials from the United States including thorium-232, radium-226, krypton-85 and tritium.)

Production of radioisotopes in Britain

The major source of radioisotopes in Britain is Amersham International. Located on the outskirts of London – Amersham is at the end of the Metropolitan tube line – the company began life as a radium refinery set up in 1940 to provide luminous paint for wartime use. Once part of the UKAEA, it was privatised in 1982 and now has thirteen subsidiaries in Europe, Japan, Australia and the United States.

The Amersham site itself is home to three cyclotrons which mostly manufacture isotopes for nuclear medicine. Their output is dominated by three radioisotopes in particular: gallium-67, indium-111 and thallium-201. For many years, Amersham also obtained radioisotopes from the UKAEA's DIDO and PLUTO reactors at Harwell, dispatched to Amersham for further processing. Since the closure of the Harwell reactors, reactor-produced isotopes must now be imported. In 1980, the company established a new site at Forest Farm near the M4 in Cardiff. The new laboratories have no reactors or cyclotrons on site but concentrate on preparing isotopes for the life sciences.

Amersham is not the only producer of radioisotopes in Britain; other research reactors and cyclotrons are also used for isotope production although their output is small by comparison. ICI's Physics and Radioisotope Services Group (PRS) manufacture radioisotopes in their Triga reactor at Billingham. Many are used by the company itself although ICI also supply isotopes to the oil and gas industries in the area and to other large chemical plants.

The Universities Research Reactor at Risley supplies local customers with short-lived isotopes (such as fluorine-18 and Argon-41) for tracer techniques and flow measurements. A similar service is offered by the Scottish Universities Reactor Centre at East Kilbride.

While Amersham supply most of the longer-lived cyclotron-produced isotopes in Britain, many shorter-lived isotopes are manufactured by the Medical Research Council (MRC) in their cyclotron at Hammersmith Hospital in London. The MRC also have a cyclotron in Edinburgh and a similar service is provided by a cyclotron at Birmingham University.

Transport

Radioisotopes are moved by all modes of transport, Rail transport is used for about 1,000 deliveries each year, with radioisotopes carried in the guard's vans of passenger trains.[3] Most, however, travel by road. An NRPB survey published in 1986 identified radiography sources as accounting for the largest number of journeys – 12,000 per year at that time. Radiography sources are gamma emitters and usually transported by car or van. By comparison, the movement of all other radioisotopes totalled 9,000 journeys, approximately 10 per cent of which are to airports and seaports for export.[4] Using a variety of vans, deliveries within the London area are mostly made by Amersham's own vehicles, while longer distances are handled by Securicor.

Deliveries can involve consignments of up to 400 packages per vehicle, usually in Type A or Excepted packages. In fact, Excepted packages, weighing just 1 or 2 kgs, make up 90 per cent of all deliveries. A typical medical isotope, for example, would comprise a radioactive liquid sealed in a glass vial enclosed within a lead capsule. The capsule would be contained inside a steel can, which in turn would be carried within a cardboard box with polysterene packing.

At the other end of the spectrum are gamma emitters like cobalt-60. Although the number of such shipments is relatively small, they contain large quantities of radioactivity and require heavy, shielded Type B containers. Amersham dispatches cobalt-60 to radiotherapy facilities around Britain and to gamma irradiation units. A typical teletherapy source will be changed every three years or so, with old sources returning to Amersham in exchange for new ones. Most consignments of cobalt-60 go by road but rail transport has also been used.

Exports of radioisotopes are made by air or road and sea, depending on the distance involved. Many European destinations are reached by road with nightly deliveries using the cross-Channel ferries. However, the relatively short half life of many radioisotopes has made air transport a necessity for some products and for more distant destinations. (For example, the parent isotope in technetium generators – molybdenum-99 – has a half life of just 2.7 days). Most exports are flown out of Heathrow Airport which handles 90 per cent of all radioactive materials exported from Britain by air.[5]

Crates for carrying new fuel. The smaller ones are normally used for magnox fuel; the larger ones for AGR fuel

Radioactive materials leave Heathrow on up to thirty flights a day. Virtually all this traffic is from Amersham International; an annual trade which involves about 150,000 packages, 7,200 flights and over forty different airlines.[6] In practice, 60 per cent of deliveries are handled by just two airlines. Of these, half are carried by Lufthansa on a daily freight-only flight to Frankfurt while the rest travel on regular passenger flights operated by British Airways. Some exports travel far afield. Gallium-67, for example, is exported mainly to Japan, while indium-111 is supplied to hospitals in the United States.

Radioisotopes are also imported. For example, although Amersham International is a major exporter of technetium generators, about half of those used in Britain are obtained from the United States and France. The 1986 NRPB survey estimated that while about 8,000 technetium generators were exported from Amersham each year, another 4,000 were imported.[7]

In addition to finished products, radioactive raw materials must be imported by Amersham International for processing, a trade which has increased in recent years. For example, cobalt-60 used to be produced in BNFL's reactors at Chapelcross and in DIDO and PLUTO at Harwell.

Supplies of this isotope are now obtained mainly from Canada, also a major supplier of molybdenum-99.

An Amersham International van delivering a consignment of radioisotopes to Heathrow Airport

Tritium, for example, is obtained from the United States Department of Energy which sells about 200 grammes a year for non-nuclear weapon uses.[8] British companies buying US tritium in the fiscal year 1988 included Saunders-Roe Developments (65.13 grams), Surelite (31.12 grams) and Amersham International (8.29 grams).[9] In the following year, the US NRC approved the export of 9,250 TBq of tritium gas to Amersham for use in products including neutron sources, electronic tubes and labelled compounds. Amersham's application for NRC approval stated that products containing DOE tritium would be exported to Australia, Belgium, Denmark, Finland, Italy, Japan, the Netherlands, New Zealand, Spain, Sweden, Switzerland, West Germany, the United States and Venezuela.[10]

The international trade in tritium has been a sensitive subject because of the isotope's military potential. Amercium-241 is another imported isotope with a military significance. As an alpha emitter it is sensitive to minute traces of smoke and used in smoke detectors. It is also used in static electricity eliminators. In 1986, Amersham International purchased 1.5 kgs of the substance, a decay product of plutonium-241, from the Commissariat à l'Energie Atomique in France.[11] French supplies of the isotope are obtained from MOX fuel fabrication plant effluents and by "cleaning" old stocks of plutonium; some may therefore be a by-product of France's nuclear weapons programme.

Other imports comprise radioisotopes, such as technetium generators, which were originally exported from Amersham and are returned to Britain after use for disposal or re-activation. Imports arrive mostly by air. Deliveries into Heathrow account for about four flights a week. Radioisotopes also arrive at regional airports, including Luton, Birmingham and East Midlands.

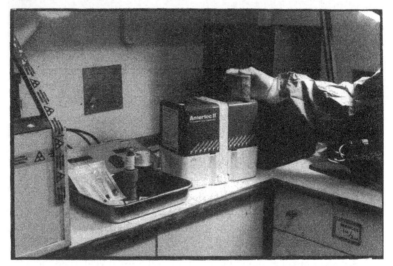

An "Amertec II" technetium generator in a London hospital. Such generators are delivered from Amersham International in a labelled cardboard box.

It is pointless trying to identify transport routes. While production is confined to a handful of sites, the diverse range of uses to which isotopes are put and the large and varying number of customers mean that isotopes can go virtually anywhere. An illustration of their ubiquity was given in 1989 when the Environment Secretary listed all the premises in Slough which were registered under the Radioactive Substances Act to keep and use radioactive materials. In addition to companies carrying out diagnostic work and non-destructive testing – predictable users of radioisotopes – the thirty-three premises also included an Asda store, W.H. Smith Do It All Ltd and even the local McDonalds.[12]

However, some traffic concentrations can be identified. With most exports dispatched from Heathrow airport, regular deliveries are made between Amersham and Heathrow via the M25. Exports are also made to Europe by road using the cross-Channel passenger ferries from Dover. Within Britain, there are regular deliveries between the laboratories at Amersham and the company's factory at Cardiff. Securicor depots in Loughborough and Bristol are also major distribution points for domestic deliveries. The specialised uses of some radioisotopes result in regular movements in other areas. For example, companies supplying well-logging services to the offshore oil and gas industries are concentrated around Great Yarmouth and Aberdeen. On a smaller scale, Heysham is a centre for similar activities in Morecambe Bay.

Because of their benefits to medicine and the fact that many radioisotopes are produced in cyclotrons – they are not therefore products of the nuclear fuel cycle – radioisotopes have not been targets of anti-nuclear criticism. Although they have had problems, these have mostly concerned their use – such as overdoses administered by radiotherapy units – rather than their transport. Nevertheless, while many radioisotopes contain only low levels of radioactivity, others are less benign and despite the attention given to fuel cycle products, radioisotopes collectively account for most radiation doses received during the normal transport of radioactive materials, most accidents occurring to radioactive materials in transit, and most of the radiation doses arising from transport accidents.

Depleted uranium

Uranium with a lower than natural proportion of U235 – "depleted uranium" – arises from two operations in the nuclear fuel cycle.

1 Reprocessing: – "Depleted reprocessed uranium" is extracted, like plutonium, during the reprocessing of irradiated fuel. The fission of uranium fuel in a nuclear reactor reduces the amount of U235; when fuel is removed from a reactor, the uranium is "depleted" in this isotope.
2 Enrichment: – The process of enriching uranium hexafluoride to obtain material with a higher than normal proportion of U-235 also creates a "tails" stream of hex with a correspondingly reduced proportion of U235, i.e. "tails depleted uranium".

In 1989 stocks totalled 29,750 tonnes, roughly two-thirds arising from enrichment operations at Capenhurst with the remainder recovered at Sellafield.[13] The stockpile is something of an embarrassment. Depleted uranium's principal value to the industry has always been as potential breeder fuel in a programme of plutonium-fuelled fast reactors. These convert depleted uranium to plutonium thereby breeding more fuel for themselves. However, with persistent uncertainties over the future of fast reactors, the major use of depleted uranium to date has been to provide new fuel for nuclear power stations; uranium recovered from reprocessing has been recycled after re-enrichment as fuel for the Advanced Gas-cooled Reactors. The material is transported in various forms between Sellafield, Springfields and Capenhurst (and Chapelcross, where some of the stockpile is stored).

The industry has encouraged other uses for the material. Its high density, high strength and high melting point make depleted uranium ideal as an alloy for special steels, for radiation shielding against gamma and X-rays, and for aircraft counterweights. (The presence of depleted uranium in the wings of the plane that crashed at Lockerbie caused some concern during rescue and clean-up operations.) Depleted uranium is also used as shielding in certain models of irradiated nuclear fuel flasks.[14]

Depleted uranium is offered for sale by BNFL as metal, dioxide, tetrafluoride and hexaflouride or alloyed with other metals like molybdenum. It is available as wire, billets, rolled plate or in a variety of cast or machined shapes. Sales of depleted uranium account for a modest export trade ranging from 30 to 150 tonnes a year.[15]

Depleted uranium offered for commercial use has a low level of radioactivity (typically just under 37 kBq/gram). Transport accidents – such as the spillage which occured at Stanstead Airport in 1977 when a package of depleted uranium ruptured when dropped – can nevertheless cause contamination, although the radiological hazards are minor.

Fuel for nuclear power stations

Three types of fuel are made in Britain for commercial (thermal) nuclear power stations. BNFL Springfields manufactures magnox and AGR fuel, which are packed in metal boxes (Type A containers) and delivered to the power stations by road. Dungeness A, for example, requires some 7,000 fuel elements each year (equivalent to 75–80 tonnes of natural uranium) while a typical AGR such as Hartlepool gets through 700–800 fuel assemblies a year (containing 30–35 tonnes of enriched uranium).[16] BNFL also fabricate fuel for pressurised water reactors including Sizewell B which will require about 64 fuel assemblies a year.[17] Approximately 130 journeys are made annually to the reactors of Nuclear Electric and Scottish Nuclear.[18] BNFL also export nuclear fuel, principally to the Tokai Mura power station in Japan. Consignments are containerised and transported by sea. (See photograph on p. 122.)

While the delivery of fuel made from natural or enriched natural uranium is a minimal radiation hazard, increasing use is being made of reprocessed uranium (RepU) which is more radioactive. The U-235 content of RepU varies depending on the type of fuel it comes from. RepU recovered from magnox fuel is depleted; recovered from enriched fuel it is depleted relative to its original enrichment, but generally with a U-235 content slightly higher than natural uranium.

Uranium recovered from reprocessing contains traces of fission products and some of the elements created by nuclear reactions. These include uranium-232 (which does not exist in nature), transuranic alpha emitters such as plutonium and neptunium, and gamma-emitting fission products like ruthenium-106. Uranium-232 in particular is hazardous because although it is an alpha emitter, its daughter products emit gamma radiation. The presence of U-232 increases the radioactivity in RepU sufficiently to require more shielding and remote handling.

Although reprocessed uranium has already been recycled extensively in Britain, output will increase in the 1990s when THORP comes on stream and recovers uranium from oxide fuels. As reprocessed uranium is recycled – through conversion, enrichment and fuel fabrication plants – its greater radioactivity marginally increases the hazards associated with the transport of the fuel materials into which it is made, for example hex and uranium dioxide.

11   Radioactive waste

If there's one part of the nuclear industry with a guaranteed future, it's waste disposal. The widespread use of radioactive materials – from industry to medicine, in power stations and weapons – generates an equally wide range of radioactive wastes. Some are more dangerous than others while some are considered safe enough to be discharged directly into the environment. In some cases, radioactive gases may be vented to the atmosphere while radioactive liquids are flushed into drains, rivers and the sea. In 1989, for example, some 808 premises in England were authorised to discharge liquid radioactive wastes into drainage systems including public sewers.[1] Many wastes, however, cannot be discharged in this manner, either because they are solid, too dangerous to release, or both. Such materials, known as "contained wastes", must be isolated in varying degrees from the environment.

Waste disposal depends on how radioactive the waste is. Nuclear reactors, for example, produce highly toxic fission products so dangerous that they must be isolated from the environment for centuries and well beyond. Harwell reportedly contains a vessel surrounded by 18 feet of concrete which must not be opened for a million years.[2] By contrast, the least dangerous items – like a broken luminous watch – can be thrown away without any special restriction. In between lies a wide variety of material which, in a controlled way and subject to various restrictions, requires disposal at authorised sites. The wide variety of wastes involved and the limited number of disposal sites – relative to a far larger number of producers – mean that the transport of radioactive waste is a bulky and burgeoning business.

For convenience radioactive wastes can be classified into four general categories with correspondingly different transport requirements. The standard classification is defined by levels of radioactivity:

Very low level wastes

These are the least radioactive wastes, which can be disposed of and transported without any special control. Sometimes referred to as "dustbin disposals", this category of waste is defined by an upper limit: beta/gamma activity must be less than 400 kBq/tonne and be less than $0.1m^3$ in volume or have less than 40 kBq of beta/gamma activity for any single item. Radioactivity at this low level is assumed not to present any significant danger.

Low level wastes

Wastes which are more radioactive than "dustbin disposals" but with an alpha content no greater than 4 GBq/tonne and a beta/gamma content no greater than 12 GBq/tonne. Typical examples include contaminated items such as used clothing, gloves, tools and equipment. This category includes a lot of wastes arising from the use of radioactive materials in research, medicine and industry: examples would include contaminated building rubble from the demolition of a luminising works, or residues containing, for example, depleted uranium, used as a catalyst in the chemical industry.

High level wastes

The intensely radioactive fission products, created inside nuclear reactors by the fission of nuclear fuel and separated by reprocessing. They are left over in the form of an acid liquor after the removal of uranium and plutonium. With lethal concentrations of alpha, beta and gamma activity, high level wastes are also referred to as "heat generating wastes" because their radioactive decay produces considerable heat.

Intermediate level wastes

The wide range of wastes whose activity exceeds the upper limits of low level material but which are not in the high level category. They are a by-product of both civil and military nuclear operations and are currently stored at the reactor sites, reprocessing and warhead plants, and naval dockyards where they arise. Intermediate level wastes include reactor components, filters and chemical residues, pieces of irradiated fuel cladding and materials contaminated with plutonium.

At present, the movement of separated radioactive waste (other than "dustbin disposals") is limited mainly to solid low level material. The most active are transported by road and rail to Drigg, Britain's major radioactive waste dump near Sellafield. Several other locations in Britain are also used for low-level disposals while less radioactive material – known as "special precautions disposals" – may, subject to DoE approval, be incinerated or buried at suitable landfill sites.

The daily train of low level radioactive waste from Sellafield, pulls off the main line into BNFL's sidings at Drigg.

With current transport operations limited to a wide range of low level materials, consistent patterns of movement or regular transport routes are difficult to identify. With over 5,000 premises in Britain registered to use radioactive substances, radioactive wastes can arise at numerous locations around the country in addition to the well-known sites of the nuclear industry.[3] Moreover, with the exception of deliveries to Drigg from the major civil nuclear establishments, there is no published information regarding the quantities of radioactive waste created by other sources - medical, industrial and, until recently, defence – nor about the quantities disposed of at locations other than Drigg. However, a brief description of current disposal practices gives a general picture of how extensive the transport of radioactive waste has become.

Current disposal of radioactive waste – special precautions disposals

"Special precautions disposals" are controlled under the 1960 Radioactive Substances Act and require permission from the Department of the Environment. Approval is given in the form of an "authorisation", usually for a site selected by the DoE. Authorisations may be subject to special conditions, and can be issued for either a single consignment of waste or for disposal at a particular site over a specified period of time.

Most of the sites are ordinary landfill tips which otherwise receive non-radioactive material. Some are local authority landfill sites; others are privately-owned. ICI, for example, dump low level radioactive waste at their Cowpen Bewley tip in Billingham. The number of sites varies over time depending on how many authorisations are in force and there is also an official reluctance to identify their locations: dumping even the lowest level radioactive material tends to provoke opposition – which may be the only time that the locations are revealed. Examples include the Walton Arpley landfill site near Warrington – which takes low-level radioactive wastes from Liverpool University and the Royal Liverpool hospital – and Cardiff city council's waste disposal sites at Ferry Road and Lamby Way, which receive radioactive waste from Amersham International's Cardiff laboratories.

The low levels of radioactivity disposed of at landfill sites warrant few extra precautions. It is normally a condition of disposal that material is disposed of in sealed plastic or multilayered sacks and delivered in closed metal bins. As for transport, any old vehicle will do. The most important restriction is the upper limit on the level of radioactivity.

Very low level wastes are also disposed of by incineration, at facilities run by either local authorities or companies such as Rechem. Low level waste which exceeds the level for special precautions disposal must be disposed of elsewhere, and that is normally to Drigg – Britain's major waste dump for the more active low level waste.

Current disposal of radioactive waste – Drigg

The sign at the entrance – "Drigg Depot" – makes no reference to the radioactive nature of what goes there. Surrounded by dense conifers, a security fence and AEA police who patrol in between, there is little to suggest that BNFL's 300-acre site on the Cumbrian coast is Britain's largest (by area) radioactive waste dump. Since 1959, Drigg has been receiving the more active types of low level waste with average annual deliveries (1983–1988) running at 37,000 cubic metres a year.[4] Much of this arrives from the nuclear power industry. Consignments typically comprise contaminated items such as paper towels and tissues, plastic bags and bottles, scrap metal, building rubble, protective clothing and soil – in fact, the everyday garbage of any industry, but which in this case is radioactive. Drigg has also been used for the disposal of plutonium-contaminated materials.

Unlike disposals at other landfill sites, the quantities of waste delivered to Drigg are published each year by the Department of the Environment. By far the largest quantities arrive from nearby Sellafield – 26,711 m³ in 1988.[5] Until 1983, all this material travelled by road, However, the regular traffic of lorries plastered with radiation symbols attracted criticism in the local villages through which they passed. After pressure from residents, BNFL built a rail link into the Drigg site which branches off the main line just north of Drigg station. The new sidings were inaugurated in 1983, and virtually all deliveries from Sellafield have since been made by rail. Although the journey is short – a distance of only 2.5 miles – the rail link can handle between 12,000 and 15,000 tonnes of waste each year – the equivalent of 10 lorry loads per day. The train service from Sellafield is currently the only regular use of rail transport for deliveries to Drigg although occasional trains have in the past delivered low level waste from other nuclear sites, for example, Trawsfynydd nuclear power station.[6]

Low level waste is transported to Drigg in skips on BNFL's own wagons, Location: Seascale, Cumbria

All other deliveries to Drigg arrive by road, principally from nuclear power stations, the UKAEA, and other BNFL sites. More material arrives from Amersham International and the Ministry of Defence, The weapons factories – especially Aldermaston – are major sources while smaller quantities arrive from the naval dockyards.

Drigg also receives the more active types of low level waste produced outside the nuclear industry. If other users of radioactive material need to dispose of waste too radioactive for a landfill site, it can be sent to Drigg through the National Disposal Service (NDS). This accounts for some 13,000 m³ of waste per year[7] and is operated by BNFL on behalf of the Secretary of State for the Environment. Waste disposed of through the NDS is either transported directly to Drigg or first delivered to Harwell and then forwarded to Cumbria. The NDS is also used for some radioactive waste arising at MoD sites.

Low level radioactive waste from Springfields passes through Preston towards the M6 on its way to Drigg

In the past, waste transported to Drigg by road was packaged into drums and carried in tipper trucks. More recently, freight containers have been introduced which are also used for final disposal. Two types of container are employed: a full-height (2.6 m x 2.4 m x 6.1 m) model for the transport and storage of 200-litre drums of waste, and a half-height (1.2 m x 2.4 m x 6.1 m) version for higher density loose material such as building rubble. A typical nuclear power station dispatches a container of radioactive waste about once every six to eight weeks.[8]

Current disposal of radioactive waste – other low level sites

Not all of the nuclear industry's low level waste ends up at Drigg. The

UKAEA have their own disposal facilities at Dounreay while BNFL themselves use two sites in Lancashire for wastes from Springfields and Capenhurst. Although the more radioactive waste from Springfields goes to Drigg, less radioactive material is transported by lorry to Clifton Marsh. Run by the North West Water Authority, Clifton Marsh adjoins the north bank of the River Ribble, just down the road from Springfields. Since 1974, it has been used mainly for material contaminated with uranium; typically ash, scrap metal and building rubble. Clifton Marsh also receives occasional consignments of waste from Capenhurst and local hospitals.

The other site is a disused clay pit at Ulnes Walton, near Leyland. Low level radioactive waste has been dumped here intermittently since 1964. Like Clifton Marsh, it has been used for the disposal of uranium-contaminated material although the quantities involved have been much less. While both Springfields and Capenhurst are authorised to use Ulnes Walton for radioactive waste disposal, the site is nowadays used only by Springfields – and very rarely at that.

Other low level wastes arise at various MoD sites, at the naval dockyards and at Rolls-Royce's submarine fuel plant in Derby. Many of these establishments, such as Aldermaston, either store waste on site or send the material to Drigg. In the case of Rolls-Royce, low level waste from their Raynesway plant is transported either to the local municipal incinerator or to Hilts Quarry, about 10 miles north of Derby. Hilts Quarry, off the A6 near Crich, is owned by Rolls-Royce and received permission for low level radioactive waste disposal in 1965.

While the vast majority of radioactive waste in transit is solid, there are also movements of liquid wastes, although the quantities and radioactivity involved are far smaller. Two examples might be noted, if only because the physical form of the material would be more vulnerable in a transport accident. Low level radioactive liquid from Capenhurst is moved by road tanker to a sewage pumping station at Meols, on the north coast of the Wirral. The liquid – mildly contaminated with uranium and technetium-99 – is mixed with domestic sewage and discharged into Liverpool Bay. Low level liquid wastes also arise at nuclear power stations in the form of radioactive oil. This is collected and transported to conventional oil-fired power stations to be burnt as fuel.

Disposal, dumping and demos . . .

Britain's nuclear past has guaranteed an output of radioactive waste well into the forseeable future. Indeed, regardless of what happens to the rest of the industry, waste transport is set to expand in new directions as more dangerous types of waste require disposal. Sooner or later, the growing stocks of intermediate and high level material – unsuitable for Drigg and currently stored at the sites where they arise – will be moved to new repositories for final disposal. This will include exports of wastes arising from the reprocessing of spent fuel for overseas customers. If past experience is anything to go by, these exports are likely to prove controversial.

Much radioactive waste is containerised for transport. This consignment (hauled by the same UKAEA lorry pictured on the cover) was heading north on the M6, presumably to Drigg.

Opposition in the past to the transport and disposal of radioactive waste is one of the reasons why such material is still awaiting final disposal. Intermediate level waste in particular has been a problem. Until 1983, much of it was dumped at sea; first in the English Channel, later in the Atlantic. The annual dumps generated a flurry of transport activity on land. In their final years, each operation typically involved moving up to 2,700 tonnes of waste, packed into drums, from various nuclear sites to the port of embarkation – over the years, British radioactive waste sailed to sea from Gosport, Newhaven, Portsmouth, Rosyth, Sheerness and Swansea. For the last dump in 1982, nine train loads of waste converged on the Gloucestershire port of Sharpness.

By the late 1970s, however. the practice of tipping radioactive waste in the sea was attracting widespread criticism. Experience was beginning to show that wastes could leak under water. In 1976, for example, the US government's Environmental Protection Agency surveyed sites where low level radioactive material had been dumped off the US coasts and found that drums of waste had corroded and in some cases broken. Moreover, samples of mud and sand removed from the same sites contained traces of plutonium and caesium, toxic radioactive contaminants which had presumably leaked from the waste despite its encapsulation in concrete and bitumen.[9] In an area of deep ocean currents, it was not inconceivable that radioactivity could disperse from the site and return to humans through the food chain. Britain's own experience of sea dumping has shown that waste can return even more quickly. During the 1964 dump, for example, material tipped overboard – which should have sunk forever – floated back to the surface in the middle of a Spanish fishing fleet.

By the 1980s, opposition to sea dumping had become an international *cause célèbre* with demonstrations against the transport operations involved. In 1980, for example, direct action by members of a British anti-nuclear umbrella group, the Severnside Alliance, stopped one of several trains transporting waste to Sharpness. Packed into barrels and loaded on open wagons, the waste had travelled from Didcot, close to Harwell and Aldermaston. Waiting at the docks was the motor vessel *Gem*, preparing to sail for the Atlantic. Conveniently for the protestors, Sharpness docks lie at the end of a short freight-only branch line which once continued across the River Severn to Lydney until a shipping accident destroyed the bridge. Nowadays, the truncated line's few trains include twice-weekly departures of spent nuclear fuel flasks to Sellafield from a siding near Berkeley power station. On Tuesday 8 July, and with no other rail traffic to worry about, the single-track line provided an ideal location for a blockade. As the waste train rounded a curve, it was stopped by demonstrators waving red flags and a 20-foot tower of scaffolding straddling the single track line. The protesters, some sitting on top of the obstruction, held up the train for over three hours. Other actions targetted the ships themselves. Three weeks earlier, Belgian protesters in Zeebrugge had boarded the freighter *Andrea Smits* in an unsuccessful attempt to prevent it sailing for the Atlantic with more radioactive waste. In the following year, Greenpeace weighed in against a British dumping operation, sailing inflatable dingies close to the *Gem* as it dropped barrels of waste overboard.

The upsurge of direct action was accompanied by diplomatic pressure from governments opposed to sea dumping. A 1983 Spanish proposal for a two-year moratorium was adopted by the London Dumping Convention (LDC), a voluntary treaty organisation set up after a UN Conference on the Environment to promote the prevention or control of marine pollution. Britain, despite its membership of the LDC, voted against the moratorium and the nuclear industry continued to plan for business as usual. However,

Protestors on a scaffolding tower halt a train carrying barrels of intermediate level waste to Sharpness docks, July 1980

athough the waste duly arrived at Sharpness, as it had done in previous years, the National Union of Seamen – whose members were required to crew the ship – imposed a boycott and scuppered the industry's plans. With their action supported by other transport unions and endorsed by the TUC, the 1983 sea dump never left the train. After an embarrassed sojourn in various railway sidings – first at Didcot, then for several months at army depots at Bicester and Thatcham – the waste returned by road to Harwell and Aldermaston.

The fiasco forced a reappraisal of waste disposal policy. Sea dumping was suspended – although the government has refused to rule it out as an option for the future.[10] Meanwhile waste continues to accumulate at existing sites with finite capacities. Short-term measures to deal with ever increasing quantities include new buildings at Harwell (storing amongst other things material from the aborted 1983 sea dump) and more efficient waste management at Drigg; by compacting waste before disposal and constructing new vaults, the life of the site has been extended to at least 2050. At Sellafield, new facilities are now in operation to encapsulate intermediate level waste and vitrify high level waste. Eventually, however, additional facilities will be needed for waste disposal on or beneath the land. Their location will influence waste transport.

In 1982, the responsibility for developing and managing new facilities for the storage of solid low- and intermediate-level waste was given to a new organisation, Nirex. Formed by the CEGB, SSEB, BNFL and the UKAEA, Nirex is responsible for disposing of waste from the nuclear industry, other users of radioactive materials – hospitals, industry and research – and from the Ministry of Defence. Nirex investigated several potential sites for different sorts of waste: at Billingham, Elstow, Bradwell, Fulbeck and Killingholme. All were abandoned after opposition and Nirex opted instead for a single national repository for low and intermediate level waste. After test drilling at the Dounreay and Sellafield sites, the latter has emerged as the industry's preference.

Apart from its ownership by BNFL, transport considerations also favour Sellafield. The proposed repository would receive some 25,000-40,000 m^3 of low level waste and 15,000 m^3 of intermediate level waste each year from the sites where the wastes arise. If all this material were transported by rail, that is equivalent to about 15 trains a week. The use of road transport for lighter packages could require up to 100 lorry deliveries each week, although up to 5 less trains per week would be needed.[11] As about 50 per cent of Britain's radioactive waste is either produced or stored at Sellafield, transport requirements will be substantially reduced if it is also the site of the repository.

High level wastes

Responsibility for managing the more dangerous but relatively smaller volume of high level waste remains with BNFL and the UKAEA. Attempts to investigate potential permanent disposal sites were abandoned in 1981 after public opposition and government policy is now to store British HLW for at least 50 years at the sites where it arises – namely, Sellafield and Dounreay. Any possibility of transport to a separate final repository in Britain has receded into the next century.

However, high level and other radioactive wastes also arise from reprocessing spent fuel for overseas customers. Since 1976, all foreign reprocessing contracts have contained an option for the return of the resulting wastes and the government have made it clear that this option should be exercised.[12] The amounts involved could be substantial. In 1988, the government estimated that BNFL's reprocessing contracts with Italy, Japan, the Netherlands, Switzerland and West Germany would create the following quantities of returnable waste:[13]

High level waste	200m³
Intermediate level waste	3,500m³
Low level waste	23,000m³

By 1990, estimates of high level waste requiring return had risen to 300m³, with the first exports expected from about 1994.[14] During the first ten years, some 2,000 containers will be exported with about 60 per cent going to Japan and the rest to Europe.[15] The total volume, however, is likely to be higher than 300m³, because the government has been considering the possibility of substituting high level waste for lower level material.[16] The problem is understandable: transporting the far larger quantities of intermediate and low level material back to Europe or Japan is not an attractive proposition: it would be much easier and cheaper to return the same quantity of radioactivity in the form of additional high level waste – known as the "equivalence concept" – which would constitute but a fraction of the volume and require far fewer shipments. The lower level wastes would remain, further justifying Britain's reputation as the infamous "nuclear dustbin".

Radioactive waste transport and public concern

The industry itself recognises that the disposal of radioactive waste is now one of the main reasons for public concern about nuclear power. With new types of waste moving around, (in new types of containers) and

new transport routes, waste transport could become a major issue raising
a predictable chorus of questions: how safe is the normal transport of such
material and what would happen in an accident?

In practice, the potential hazards will vary considerably because of the
range of material involved. A broad distinction might be made between
wastes which require shielding in transit – intermediate and high level
wastes – and those, like low level wastes, which do not. Despite their
shielding, transport containers loaded with intermediate or high level waste
will still emit a small amount of gamma radiation. For intermediate level
waste an NRPB report notes that ". . . certain individuals could theoretically
receive significant doses."[17]

Potential dangers could be increased by an accident, especially as some
future consignments of waste will contain large quantities of radioactivity.
On the face of it, there is little to worry about. The safety potential of
intermediate and high level wastes will be enhanced by the physical form
of the material: high level waste will be vitrified and most intermediate level
waste encapsulated in concrete or a polymer. Solidification should ensure
that much of the alpha and beta radiation will be sealed inside the waste,
even if an accident deprived it of shielding. If a transport container were
ruptured, immobilised waste should not disperse into the environment.

Nevertheless, the industry's own analyses show that an accident could
still release radiation. A leak from a consignment of intermediate level waste
at Willesden Junction in London formed the subject of the second part of
the NRPB's report mentioned above. This hypothetical accident involved a
consignment of magnox silo sludges, the sort of material that will eventually
be transported from power stations like Bradwell and Dungeness to a new
repository. The NRPB's scenario assumed that the 3.5 m^3 consignment of
waste included the quantities of "radiologically significant" radionuclides
outlined in Table 11.1.

Using the MARC computer programme, the NRPB considered an
accident which released a plume of radioactivity in respirable form,
escaping continuously for ninety minutes. Public exposure to radiation
would arise from the inhalation of airborne material and by external
irradiation from airborne and deposited material. As the study assumed that
only a tiny fraction of the total contents were released (10^{-6} to be precise)
the predicted health effects were correspondingly small. With such a small
release of radioactivity the average individual risk of fatal cancer would be
minimal; at worst, one in 5,000,000,000 at a distance of 100 metres from
the leak.

This assessment was for intermediate level waste. Larger amounts of
radioactivity will be contained in consignments of high level waste.
Each container of vitrified material produced at Sellafield will typically
contain some 40,000 TBq of radioactivity (at the time of production).[18]

With high level waste transport flasks holding about such 20 containers, the total inventory of radioactivity – 800,000 TBq – will approach that of a consignment of spent fuel. This alone is likely to generate concern. However, while flasks of spent fuel include radioactive gases and liquids, vitrified waste is solid. Arguably, if vitrified waste can survive the vicissitudes of of geological time, it ought to be proof against transport accidents. Although loss of shielding would increase gamma emissions, its solidity should prevent it dispersing into the environment.

Table 11.1: Contents of a consignment of intermediate level waste

Radionuclide	Radioactivity (GBq)
Cobalt-60	18,800
Krypton-85	3,110
Strontium-90	50,300
Yttrium-905	0,300
Ruthenium-106	2,010
Caesium-134	1,130
Caesium-137	69,000
Cerium-144	1,980
Plutonium-238	154
Plutonium-239	1,000
Plutonium-240	1,010
Plutonium-241	38,900
Americium-241	1,630
Curium-242	8.1
Curium-244	13.2

Source: NRPB

Shipments of low level waste could also attract controversy. Unlike more radioactive material, low level waste will not be transported in specially designed flasks, but in drums and freight containers. Although it may be compacted before transport, low level waste is not encapsulated or vitrified into a solid block, but packed in the physical form in which it arises. Furthermore, such wastes contain contaminated items made from plastic, rubber, paper and cloth, which are therefore combustible. Thus of all the different types of waste, low level material will be the most vulnerable to the effects of fire or a crash: it takes relatively little force to rupture a 200-litre drum or a standard ISO freight container. Although the contained radioactivity is small, low level waste will be transported with far less protection than most radioactive materials and in the event of a serious accident, could be dispersed into the environment. As past experience at Drigg has shown, when radioactive waste catches fire, radioactivity blows downwind.

Moreover, low level waste is defined not by the presence of particular radionuclides, but by their concentration in the waste material – even low level waste may contain small amounts of substances like plutonium. The presence of plutonium has been a source of confusion, even regarding existing waste disposal. In 1976, for example, the Flowers Report noted that Drigg contained a third of a tonne of plutonium, the cumulative result of burying plutonium-contaminated waste at the site.[19] By 1988, the government revealed – quoting BNFL as its source – that the quantity had risen to half a tonne.[20] Yet in a television programme broadcast in 1985, BNFL's technical director, Dr Bill Wilkinson, publicly denied that was there was any plutonium at Drigg.[21]

Such confusion about the content of low level waste undermines reassurances about transport safety. The presence of substances like plutonium, even in minute amounts, has implications for transport, especially if a consignment of waste was involved in an accident: a spillage or fire could end up dispersing plutonium-contaminated material into the environment, despite the ostensibly innocuous nature of low level waste.

In any case, regardless of what a consignment of waste is *supposed* to contain, there is always the possibility of a mistake. Errors have already affected arrivals at Drigg, where deliveries of low level waste have been found to be more radioactive than they should have been. One culprit was recorded by BNFL in 1984: "A quantity of waste sent to the Drigg disposal site . . . exceeded the limit for burial as a resut of incorrect information provided by Harwell."[22] In normal circumstances, such mistakes would have little or no transport significance. But they show that errors *can* be made whose consequences could be worse: any increase in the gamma activity of waste is a potential increase in the dose to transport workers and possibly to members of the public. An increase in any type of radioactivity could increase hazards if an accident occurred.

Box 11.1: New transport containers

Anticipating a growth in the movement of radioactive waste, Nirex are developing a standardised range of transport packages for low and intermediate level solid wastes. Three types of package have been developed for each type of waste.

Most low level waste will travel in 200-litre steel drums transported inside a reusable freight container. Larger items will utilise either a $3m^3$ steel box or a $12m^3$ steel or concrete box. The latter is intended mainly for decommissioning wastes.

Box 11.1 (continued)

Most intermediate level wastes will require a 500-litre steel drum for transport. As with low level wastes, a $3m^3$ steel box and a $12m^3$ box will be available for wastes which will not fit into a drum. The larger box will be a steel-clad reinforced concrete structure with a wall thickness determined by shielding requirements. Both the 500-litre drum and the $3m^3$ box will be transported in a specially developed reusable transport container providing further radiological protection. The new container will hold either 4 drums or one box and have a wall thickness ranging from 70 to 300 mm as required. Nirex anticipate a fleet of over 300 such containers.

New containers will also be required for the transport of high level waste. This contains many of the same radioactive substances which make spent fuel so dangerous. (The major difference is that high level waste contains no fissile materials – uranium and plutonium are removed by reprocessing – and package designs do not need to be proof against criticality.) The design of a suitable container has been undertaken jointly by BNFL and NTL. The container must be robust enough to withstand a major accident or fire and incorporate substantial shielding to reduce the emission of gamma radiation. It must also allow internal heat to disperse. The end product is broadly similar to a cylindrical spent fuel flask, although the contents will not be fuel elements but the output of Sellafield's vitrification plant: sealed stainless steel containers (approximately 1.3 metres high with an outside diameter of about 400 mm) holding about 400 kgs of high level waste. Each flask will carry some 20 cylinders of vitrified material and have a total laden weight of just over 100 tonnes.

The transport arrangements for high level waste will mirror those for imports of spent fuel: the wastes will be exported by the same land and sea routes which bring foreign fuel to Britain for reprocessing. Most flasks will be returned in BNFL and PNTL's purpose-built ships, which ply between Britain, Europe and Japan. Within Britain, exports of waste will be carried by rail on the same wagons used for imports of foreign spent fuel. The routes will also be the same, except that waste flasks will travel in the opposite direction. Given the predominance of Japanese contracts in THORP's order book, much of this new traffic will be concentrated on the short stretch of line between Sellafield and Barrow. However, high level waste must also be returned to BNFL customers in Europe and if the same ports are used as those through which European spent fuel currently arrives, that will mean lengthy journeys across England – notably south through London to Dover.

12   The military role of the
        British nuclear industry

A few years ago, an unmarked military vehicle broke down on the A4 near the centre of Newbury. Although caused by nothing more serious than a mechanical fault, the incident, which happened in the morning rush-hour and held up the traffic, made the front page of the local newspaper. The reason? The vehicle, accompanied by a police escort, had been transporting nuclear material – reported to be plutonium – from Aldermaston. In the event, there was neither a fire nor a crash, and no one was injured. It was, after all, only a breakdown. Although the Ministry of Defence did not confirm the contents of the vehicle, it reassured the newspaper that the public had never been endangered, noting that: "These are specially-manufactured vehicles designed for the carrying of this sort of material".[1]

In conclusion, a trivial incident, but one which also revealed something else – the vehicle's destination: the nuclear material had been on its way to Harwell. In other words, it was travelling between a military establishment and one usually presented as a civil site; Harwell is officially known as the Atomic Energy Research Establishment. Such movements constitute the physical links between the civil and military sides of the nuclear industry. Despite attempts to separate the two, they are inextricably linked.

Between them, nuclear warheads and submarine reactors require plutonium, highly enriched uranium, depleted uranium and tritium, none of which are made by the Ministry of Defence. Rather, they are products of the same technical processes and many of the same nuclear facilities – principally the factories of British Nuclear Fuels – which produce fuel for nuclear power stations. Military establishments like Aldermaston, Cardiff and Burghfield do not function in self-contained isolation, but as an integral part of the same nuclear industry which fuels the likes of Sizewell B.

As a result, many radioactive materials – from uranium ore to plutonium, via intermediate products like hex and spent fuel – have a dual purpose, helping both civil and military programmes. Such considerations have implications for transport: any consignment of radioactive material which might have a military use is a campaigning target for those opposed to nuclear weapons: its military potential is an issue in itself. A brief review of the nuclear materials required for military use shows how the industry's civil and military activities have overlapped to each other's advantage.

Plutonium

Plutonium is produced automatically in all working reactors, and there is a popular belief that, wherever it is found – in fast reactors, exports, or spent fuel – it could end up in nuclear weapons. While it is officially acknowledged that plutonium from Calder Hall and Chapelcross can be used for military purposes, the possibility that plutonium produced by other British nuclear power stations might be put to military use – thereby giving spent fuel transport a military purpose – has been a recurring suspicion.

It is complicated by a technicality: the difference between weapons-grade and reactor-grade plutonium. The irradiation of nuclear fuel inside a reactor produces several different isotopes of plutonium. Most is plutonium-239, although the proportion of other plutonium isotopes increases over time as the fuel remains in the reactor. Nuclear warheads, however, require plutonium with the highest possible proportion – weapons-grade – of plutonium-239; the greater the proportion of other isotopes, the less suitable the plutonium is for military use. Weapons-grade plutonium is therefore manufactured by removing fuel sooner than would be necessary if the reactor were optimised for electricity generation.

If this suggests that commercial nuclear power stations have little military potential, that is not necessarily the case. The distinction between reactor-grade and weapons-grade plutonium is blurred by two qualifications. First, civil reactors produce plutonium with a higher than normal percentage of plutonium-239 at the beginning and end of their lives. This occurs because a new reactor begins operating with a full load of unused fuel, some of which must be prematurely removed to enable the reactor to arrive at a "steady state" for regular refuelling. Secondly, even reactor-grade plutonium might be used in a nuclear weapon. This possibility surfaced at the 1977 Windscale Inquiry and heightened concern about the proliferation potential of spent fuel reprocessing. The point was underlined by the subsequent revelation that, although reactor-grade plutonium is far from ideal, the US had exploded a bomb made from reactor-grade plutonium.[2] More recently, the Director General of the IAEA has stated that the Agency considers any isotopic mix of plutonium (except for that containing more than 80 per cent Pu-238) to be capable of use in a nuclear explosive device.[3]

With Magnox reactors producing both more plutonium (per megawatt of generated electricity) and plutonium with a higher proportion of Pu-239 than other reactor types, there has been a widespread suspicion that plutonium from Britain's civil nuclear power stations has been used in nuclear weapons. Spent fuel, for example, would then have a military purpose. Official denials, repeated over the years, have been undermined by a discrepancy between the amount of plutonium officially accounted for and the amount actually produced by the civil nuclear programme. (Plutonium production at Calder Hall and Chapelcross has never been officially revealed because of its explicit military potential.) Independent calculations of total civil production have concluded that approximately 6–6.5 tonnes of plutonium has gone "missing".[4] Part of this discrepancy is officially acknowledged in annually published accounts which quantify plutonium stocks: the figures always exclude material exported to the United States prior to 1971. These undisclosed quantities were exported as a result of the 1958–9 Mutual Defence agreements, under which nuclear materials were exchanged across the Atlantic (US supplies of highly enriched uranium and tritium were swapped for British plutonium). As this was an explicitly military agreement, might this material have gone into US nuclear weapons?

The British government has repeatedly asserted that plutonium sent to the US from Britain's civil reactors has never been used for military purposes.[5] Even though it was exchanged under a military agreement, most of it has supposedly ended up in the US fast reactor programme. Nevertheless, the government has always refused to reveal the amount exported because of "the barter arrangements under which plutonium was consigned"; the secrecy is designed to conceal how much military nuclear material Britain has received in return. In 1984, however, an anticipated shortage of plutonium in the United States raised questions about Britain's involvement which yielded several revelations. First, the US Energy Secretary Donald Hodel confirmed that imports of plutonium during the 1970s from the British military stockpile had indeed been used in US warheads. Second, although the US administration ackowledged that none of Britain's civil plutonium exports had at that stage been put to military use, it refused to rule that option out for the future. As Hodel made clear, the civil plutonium imports had been acquired under the 1958 Mutual Defence agreement which "permits the use of any plutonium obtained thereby for defence purposes".[6]

Finally, according to the *International Herald Tribune*, the United States had received some 4 tonnes of plutonium from Britain's civil Magnox reactors.[7] This statistic gives another reason why the British government has never revealed such information: a comparison with the discrepancy between estimated plutonium production from the civil Magnox reactors and the amount of material officially accounted for – a discrepancy of some 6–6.5 tonnes – still leaves approximately two tonnes of civil plutonium missing. While an estimated 1.2 tonnes is now acknowledged to be present in waste stored at Sellafield[8], it might be assumed that the rest has gone to the Ministry of Defence.

Needless to say, military usage has been repeatedly denied by government ministers. In 1983 for example, Energy Secretary John Moore categorically stated: "No plutonium produced in any of the CEGB's nuclear power stations has ever been used for military purposes in this country . . . Further, no plutonium from the CEGB nuclear programme has ever been exported for use in weapons."[9] Even the Prime Minister implied as much in 1985: "Plutonium from CEGB and SSEB power stations is not used for nuclear weapons".[10]

Such statements were echoed by the CEGB in their evidence to the Sizewell Inquiry – until, that is, the Board's former chair, Lord Hinton, branded their evidence as "bloody lies".[11] His blunt accusation led the CEGB's chair, Lord Marshall, to contradict the denials of successive governments and publicly admit for the first time what was already suspected: plutonium from the CEGB's early reactors had indeed gone into the defence stockpile.[12] One year later, the Prime Minister tacitly admitted as much in a carefully worded parliamentary answer: "No plutonium produced in civil reactors, in this country has been transferred to defence use or exported for such use during the period of this administration".[13] If the implications of that weren't obvious enough, a statement three days later was even more incriminating: "I am not able to answer for previous Administrations".[14]

After years of obfuscation, it is now clear that plutonium from the civil nuclear power stations has been put to military use. The one remaining uncertainty – apart from the quantities involved – is whether that use was British or American or both: if civil plutonium went into the British military stockpile, and recalling US Energy Secretary Hodel's confirmation that plutonium from the British military stockpile had been used in US warheads, then the civil nuclear reactors of the CEGB and SSEB could well have armed US warheads. If that was the case, it raises the possibility that parts of the warheads of, say, US bombers now based in East Anglia, may, 20 years or so earlier, have been trundling down the railway line from Sizewell in a spent fuel flask on its way to the Windscale reprocessing plant.

Highly enriched uranium

In the days when Britain produced its own supplies of highly enriched uranium at Capenhurst, it was the end product of a convoluted military fuel cycle which integrated all of the sites now run by BNFL: hex from Springfields was transported to Capenhurst for high enrichment; Capenhurst's output was reduced to highly enriched uranium metal at

Windscale from where it was delivered to Aldermaston[15] Although British production ended after the 1959 Mutual Defence Agreement, the weapons-grade uranium has been re-used in successive generations of British warheads and continues to benefit the Ministry of Defence. Capenhurst's output during the 1950s and 1960s has been incorporated, for example, in the Chevaline warheads for Polaris submarines.[16]

Since the closure of the high enrichment plant at Capenhurst in the early 1960s, Britain has relied on the United States for supplies, principally for submarine fuel. Supply arrangements have nevertheless been helped by the civil nuclear programme. This was acknowledged in 1983 by Energy Secretary John Moore who, despite his previous assertion (quoted above) that "no plutonium produced in any of the CEGB's nuclear power stations has ever been used for military purposes in this country . . ." also conceded that "the export of plutonium (from civil nuclear power stations) has . . . benefited the UK defence programme" – the reason being the highly enriched uranium that Britain received in return.[17]

Dounreay has also had military connections. Although better known for its work with plutonium, the UKAEA site has manufactured highly-enriched uranium fuel for the MoD's research reactors. Dounreay's fuel fabrication plant has supplied the highly enriched uranium fuel for Aldermaston's Herald reactor (now closed)[18] and for the Jason reactor at Greenwich.[19] Irradiated fuel from the two reactors has also been reprocessed by Dounreay.

Dounreay's activities even have a potential, albeit remote, connection with US nuclear warheads: Dounreay has supplied highly enriched uranium fuel to research reactors overseas which may have been returned to the United States for use in the US nuclear weapons programme. This possibility stems from Dounreay's use of US-supplied highly enriched uranium. After the closure of Britain's high enrichment plant in the early 1960s, the research reactor fuel manufacturing plant obtained fresh supplies of highly enriched material from the United States. Over the years, new fuel has been supplied to research reactors in countries including Australia, Denmark, West Germany and South Africa. While some exported fuel has been returned and reprocessed at Dounreay, other fuel has been reprocessed in the US. The military connection surfaced in 1985 when the US General Accounting Office (GAO) revealed that highly enriched uranium fuel of US origin, returned to the US from foreign research reactors, was being reprocessed by the US and the recovered uranium re-used as "driver fuel" in US military production reactors.[20] Driver fuel is used to irradiate a blanket of uranium and lithium in which plutonium and tritium are manufactured for US nuclear warheads. Such products are not extracted from the driver fuel itself. Coincidentally, the countries which have returned research reactor fuel to the US for reprocessing include Denmark, West Germany and South

Africa.[21] As all these countries have in the past received research reactor fuel made at Dounreay, the UKAEA's plant may have unwittingly played a small but no doubt useful part in the US weapons programme.

Tritium

Tritium, a radioactive isotope of hydrogen, provides a fusion reaction in thermonuclear warheads. It is manufactured by irradiating the non-radioactive isotope lithium-6. Tritium production comprises two stages: separating the isotope lithium-6 (which constitutes only 7.42 per cent of naturally-occuring lithium, the remainder being lithium-7), and irradiating lithium-6 "target elements" inside a reactor.

Like the fissile ingredients of nuclear warheads, tritium has been manufactured by British Nuclear Fuels and its predecessor, the UKAEA. In the late 1950s, the UKAEA carried out both stages of production at Capenhurst and Windscale respectively. In 1956, a prototype lithium isotope separation plant was commissioned at Capenhurst. A larger plant followed soon afterwards.[22] Lithium-6 was irradiated first in the Windscale reactors and then at Calder Hall.

Like highly enriched uranium, domestic production of tritium ceased after the 1958/9 Mutual Defence agreements. Both materials were subsequently obtained from the United States in exchange for plutonium. In 1976, however, the Ministry of Defence placed a contract for the supply of tritium with British Nuclear Fuels.[23] A new plant was completed in 1980 at BNFL's Chapelcross Works (which otherwise produces electricity for the English National Grid) to manufacture lithium elements for irradiation in one of the site's reactors and reprocess them to extract tritium.

Depleted uranium

Depleted uranium has an ambiguous status: it arises as a by-product of civil and military reprocessing and enrichment and is used for both civil and military purposes. The size of the military stockpile has always been secret, although an MoD contracts bulletin provided a hint in 1989 by announcing an intention to store between 400 and 800 containers of depleted uranium hexafluoride "at a licensed nuclear site" for a ten-year period from October 1989. Eight hundred type 48Y hex containers would hold over 9,000 tonnes of depleted UF6.

The stockpile provides a classic example of how civil nuclear energy has been subsidised by the military nuclear programme. When BNFL was set up in 1971, it not only took over most of the nuclear fuel processing factories of the UKAEA, but also large stocks of depleted uranium, which – according to the UKAEA – "were mainly arisings from past military production. . .".[24] These stocks were given to BNFL free of charge and recycled through the Capenhurst enrichment plant. After re-enrichment, the uranium provided fuel for the electricity generating boards' Advanced Gas-cooled Reactors. The financial benefit derived by the generating boards from this military bequest was spelt out by the UKAEA's comptroller and auditor general: ". . . their use free of charge was considered essential to enable the modernised Capenhurst plant to produce enriched uranium at reasonably competitive prices, and the prices offered to the Generating Boards under the ten-year supply contracts assumed the continued availability of this material free of charge . . .".[25] Thus, by inheriting a military by-product free of charge, the generating boards – and the AGR programme in particular – have been subsidised by Britain's military nuclear programme.

Depleted uranium has a number of important military uses. Nuclear warhead "tampers" are manufactured from depleted uranium by the Ministry of Defence at Cardiff. In addition to its extremely high density, depleted uranium when finely divided is naturally pyrophoric; that is, it can spontaneously catch fire, especially on impact with metal. These qualities have made depleted uranium ideal for use in certain types of ammunition. Not only will it pierce thick armour, but it will ignite on impact, setting fire to surrounding inflammable materials like fuel or ammunition. Applications have ranged from anti-personnel uranium bullets to a variety of incendiary penetrators designed to be fired from tanks, aircraft or ships. Depleted uranium is used by the armed forces of various countries, including the US Air Force, which stores depleted uranium ammunition in Britain for its A10 aircraft.

Britain's own military interest was declared in 1979 when the Defence Secretary announced a programme of research and development for depleted uranium ammunition.[26] Test firing of depleted uranium penetrators has been caried out at the Eskmeals Proof and Experimental Establishment on the Cumbrian coast near Sellafield.[27]

Another application is the Phalanx Close-in Weapon System. Designed by the US Navy, Phalanx consists of a six-barrelled 20 mm cannon which can fire 3,000 depleted uranium bullets per minute. Several systems were hurriedly purchased by Britain in 1982 during the war with Argentina. By 1985, four Phalanx systems with depleted uranium ammunition were in use with the Royal Navy, on *HMS Invincible* and *Illustrious.*[28]

Paradoxically, the high density of depleted uranium has also been exploited for the opposite effect. Tanks such as the US Army's M1 Abrams model have been clad in a mesh of depleted uranium to provide protection *against* weapons such as anti-tank missiles.

Supplies of depleted uranium for British use are obtained from domestic and overseas sources. Britain's nuclear industry has supplied material for the penetrator research and development programme announced in 1979,[29] and it might be assumed that depleted uranium for British warheads is also derived from domestic sources. If that is the case, then supplies will most likely come from Capenhurst's stockpile of enrichment tails. This is stored in the form of hex and would require transport to Springfields for conversion to metal before being delivered to AWE in Cardiff.

Despite Britain's massive stockpile of the material, some of the MoD's requirements are met by imports.[30] These are likely to be stocks required for ammunition. (A BNFL plan to build a plant at Springfields to make depleted uranium anti-tank penetrators for the MoD fell through in 1981.) Some idea of the quantities involved came to light in 1989. Information obtained under the US Freedom of Information Act revealed imports during 1987 from three US companies, all of whom had originally obtained enrichment tails from the US Department of Energy. Eighteen consignments were supplied by Rockwell International, with a further two from Sandia Laboratories in New Mexico. A third company, Martin Marietta Energy Systems of California, dispatched 24 separate shipments, each weighing up to a tonne. Reportedly, the imports were acquired for use in the Phalanx system and in a programme of upgrading the Challenger tank.[31] Thus one of the least dangerous radioactive materials moving around makes a vital contribution to both nuclear and conventional weapons.

Past, present and future . . .

British Nuclear Fuel's contribution to Britain's nuclear deterrent is absolutely crucial. According to its annual reports, roughly 10 per cent of its business is with the Ministry of Defence. Elsewhere, however, the company's military activities are downplayed or even denied.

A classic example is provided by the management of Springfields which manufactures two indispensable military products: uranium fuel (from which weapons-grade plutonium is made) and uranium hexafluoride (used to obtain highly enriched uranium). After accusations from campaigners that Springfields processed uranium for military use, the plant's management issued an unequivocal denial to a local newspaper: ". . . the fuel we make is for power stations, not for military purposes".[32] When it was pointed out that even the government had described Springfields as a "mixed civil-military processing plant"[33] the management changed its tune: "The reason it was stated that fuel is not processed for military use at Springfields is that we make the fuel for all power stations at home, and some overseas. If the Government wants the fuel for military purposes there are two power stations which can make it. We supply the fuel for these two . . ."[34] The two, of course, being Calder Hall and Chapelcross whose production of electricity, plutonium and (at Chapelcross) tritium would grind to a halt without fuel from Springfields.

Even when BNFL do acknowledge their military work, it is presented as a completely separate activity with no connection to the civil side of the industry. Company Directors Con Allday and Dr William Wilkinson told the 1985-86 Select Committee of the Environment that the peaceful and military developments of nuclear technology were "not interdependent" but "completely separate production activities".[35] Yet the same fuel elements which produce weapons-grade plutonium at Calder Hall and Chapelcross are simultaneously generating electricity for the national grid; the same reprocessing building at Sellafield is used to extract both weapons-grade and reactor-grade plutonium; the same hex plant at Springfields manufactures uranium hexafluoride for military and civil enrichment.

Materials like hex are therefore seen to have an ambiguity. Hex can be enriched to provide nuclear material for either civil power stations or the Navy's submarine reactors. Consignments of hex are shipped across the Atlantic for both purposes. Thus even though military production accounts for only a minor share of the industry's output, the possibility that some radioactive materials will be put to military use tars many more with the same brush. Moreover, regardless of current use, most nuclear materials, from yellowcake onwards, could have a potential military application in the future – the civil materials of today could become part of tomorrow's nuclear deterrent.

Predicting the Ministry of Defence's future nuclear material requirements is inevitably speculative, related as it is to the future of the nuclear deterrent which itself depends upon domestic and international politics. However, if Britain's nuclear arsenal is to be maintained – perhaps well into the twenty-first century – those responsible for forward planning in the Ministry of Defence would have to ensure that supplies of fissile material and tritium will be adequate in years to come. If a future government decided, for example, that Trident's warheads, after a decade or so of deployment, needed modernising or "improving" – just as the Chevaline programme updated Polaris – the demand for nuclear materials might even increase.

Especially important is the availability of plutonium and tritium currently produced at Calder Hall and Chapelcross. These reactors are nearing the end of their lives and will eventually close, thus depriving the MoD of their only officially acknowledged sources of these materials. The obvious alternative is to replace them with new dual-purpose reactors, which BNFL have been actively considering. A feasibility study, revealed in 1988, has been looking at alternative reactor designs which would, according to the company, ". . . depend on MoD requirements . . ."[36] There are however, other options and speculation about future supplies of military material, especially plutonium, shows how the civil side of the industry could well have a military role.

In the past the Prototype Fast Reactor has been suggested as an ideal source of weapons-grade plutonium. The blanket of depleted uranium surrounding the core can be used to breed plutonium with a high proportion of Pu-239, although the government has denied that it will be put to military use.[37] Nevertheless, if the PFR or any other future fast reactor were used to produce military plutonium, it would cast a military shadow over many nuclear materials currently moving around. Not only would a fast reactor's fuel – plutonium and depleted uranium – be helping the Ministry of Defence, but by extension, so would substances like spent fuel and uranium hexafluoride from which plutonium and depleted uranium are obtained. It would be no exaggeration to say that if a British fast reactor were ever used for military purposes, almost the entire civil nuclear power industry would be indirectly implicated in that activity. However, cutbacks in fast reactor funding and its likely demise suggest that this option must now be ruled out.

Others, however, are still possible. Plutonium might also be obtained from civil spent fuel. There is now no longer any doubt that in the past at least, plutonium produced in Britain's civil nuclear power stations has been put to military use. During the 1960s in particular, flasks of spent fuel from the CEGB and SSEB's nuclear reactors were travelling to Sellafield partly for the benefit of the Ministry of Defence. In recent years, these military connections are supposed to have disappeared but they may arise again in the future. In particular, Britain's Magnox reactors will acquire a renewed military significance as they near the end of their lives. Just as weapons-grade plutonium is created at the beginning of a reactor's life, so closure also produces plutonium of higher than normal quality, that is, plutonium with a higher proportion of Pu-239: the last to be loaded (and therefore least "burnt") will contain the purest plutonium. The closure of the nine Magnox power stations therefore promises a glut of high quality, potentially weapons-grade, plutonium – a windfall which will surely not go unnoticed by the Ministry of Defence. If plutonium has been switched from civil to military use in the past, it could presumably happen again. Berkeley nuclear power station illustrates the potential: reactor No. 1 which shut down at the end of March 1989, had been refuelled less than twelve months before.[38]

In the longer term, there is also the possibility of producing military plutonium by laser enrichment. The technology is better known as a method of enriching uranium without the bother of feeding it through diffusion plants or centrifuges. The UKAEA have been researching laser enrichment since 1974, while BNFL expect to build a "demonstration laser enrichment module" for separating uranium in the mid-1990s. However, laser enrichment can also be used to obtain weapons-grade plutonium, an application which is being actively pursued in the United States. Laser enrichment could purify reactor-grade plutonium to weapons-grade requirements, a possibility which has reportedly interested the Ministry of Defence at Aldermaston.[39] If developed in Britain, it would enable reactor-grade plutonium from any nuclear reactor to be upgraded for use in weapons.

A final alternative is that at some point in the future the government will simply change its mind. Assertions in the present tense – ". . . the Government have no plans for putting to military use any plutonium derived from, etc. . ." – are easily nullified by time. In any case, past experience shows that statements about the military aspects of the nuclear industry too often fall short of the truth. If the Ministry of Defence needed plutonium, official secrecy would probably ensure that it could be obtained from any source and if necessary without being revealed. When something as important as the ability to wreak death and destruction on millions of foreign citizens is at stake, a minor casualty like the truth can easily be overlooked.

Links between Britain's civil nuclear industry and foreign nuclear weapons

Britain's nuclear industry not only sustains British nuclear warheads, but potentially helps nuclear weapons programmes in other countries. This international connection between civil and military nuclear power is related to the import and export of nuclear materials by companies like BNFL.

The major foreign beneficiary of Britain's nuclear industry has undoubtedly been the United States. The nuclear material exchanges, and the "benefits" derived from them by both governments, far outweigh any other bilateral trade in radioactive material. However, other trade links have also had military overtones. Although nowhere near the UK/US links in either scale or military importance, they cast more military shadows over some of the ostensibly civil activities of Britain's nuclear industry. Two other existing nuclear weapon states, France and Israel, have derived military benefits from British nuclear exports, as have several other countries, recognised as potential nuclear powers in the future.

France

As in Britain, France's nuclear industry began with Magnox-type reactors, built to produce both electricity and plutonium for nuclear weapons. The G1 reactor at Marcoule produced France's first nuclear electricity in 1956 and four years later France exploded its first nuclear bomb. Despite this apparent self-sufficiency, France has on occasion obtained supplies of plutonium from Britain.

In the mid 1960s, the UKAEA, then responsible for the Windscale reprocessing plant, delivered 90kg of plutonium oxide to France for use in "Rapsodie", an experimental fast reactor at Cadarache.[40] Although the plutonium was sold to the Supply Agency of Euratom (acting on behalf of the French) under a safeguards agreement signed in 1959, the exports indirectly benefitted France's nuclear weapons programme as the diplomatic correspondent of the Times pointed out at the time: "One somewhat unexpected aspect of the deal, however, is that without such a sale . . . the French Government either would be unable to put the reactor into operation, or would be faced with a drain on its resources of plutonium produced in France, which are required for the French military nuclear programme."[41] Although Britain's plutonium did not (presumably) end up in French nuclear warheads, the Rapsodie connection was comparable, albeit in a far smaller way, to the "civil" use of British plutonium exported to the US under the 1958/59 Defence Agreements: British supplies for civil purposes released more of the recipient's own material for military use.

More recently, France's ageing dual-purpose reactors have closed, raising questions about how will it produce plutonium for nuclear weapons. In 1982 the house magazine of the electricity generating company, Electricité de France (EdF) gave one answer – the Superphénix, the largest fast reactor in the world. Built at Creys-Malville as a collaboration between France, Belgium, Italy, Britain and West Germany, the 1,200 MWe reactor reached full power at the end of 1986. According to an article in the magazine, the reactor would produce enough plutonium to make about 60 atom bombs each year. "Under these conditions, Superphénix becomes . . . the technical basis of the French nuclear military force."[42]

Although the British government has denied all knowledge of the French possibility, it would – if true – indirectly implicate Britain in the production of French nuclear weapons. The connection, as one might expect, is BNFL which has supplied a proportion of the plutonium for Superphénix's fuel. The plutonium is not British, but material recovered by BNFL at Sellafield from reprocessed spent fuel from Italy's Magnox reactor at Latina. By 1982, over half a tonne of plutonium from Latina had been allocated to Superphénix.[43] BNFL's contribution to Superphénix will not itself end up in French nuclear warheads, but while Sellafield's material fills part of the reactor's core, France's military supplies can be bred in the surrounding "blanket" of depleted uranium. In this way, by helping to fuel a reactor used for military purposes, BNFL's foreign spent fuel reprocessing activities are tainted with military overtones.

By extension, BNFL's link with Superphénix casts a military shadow over other imports of foreign spent fuel. So far, virtually all the plutonium supplied from Sellafield to Superphénix was originally manufactured in Latina, one of only two foreign reactors whose fuel is currently reprocessed by BNFL. However, this situation will change when THORP comes on stream in the 1990s. THORP will extract plutonium from the larger quantities of oxide fuel which have been arriving in Britain from Europe and Japan. These include substantial deliveries from nuclear power stations in Germany, which as a partner in Superphénix, is, like Italy, scheduled to provide a proportion of the reactor's plutonium. Superphénix will need regular reloads of plutonium fuel which can only be obtained by reprocessing. Thus plutonium extracted by THORP from West German spent fuel could, if used in Superphénix, also aid France's military programme. For those concerned about imports of spent fuel through ports like Barrow and Dover – and especially deliveries from Italy and West Germany – the complexities of the nuclear fuel cycle and its military connotations will provide much scope for speculation and controversy.

Israel

An indirect link between British Nuclear Fuels and Israel illustrates how the civil nuclear industry can help military programmes by accident. In 1984, BNFL delivered approximately 42 tonnes of depleted uranium to a Luxembourg company (International Metals S.A.) for use in the production of alloys for steel making. The tails depleted material from Capenhurst was exported under safeguards and the company reportedly gave assurances that the material would only be used for peaceful purposes. Despite that promise, BNFL's uranium was subsequently resold to the United Steels Company – an Israeli firm which bought the material for use in anti-tank shells.[44]

Such transactions probably raise few eyebrows within the nuclear industry. It was, nevertheless, an embarrassment for Euratom and the IAEA who should have been notified of the resale but were not. It may also be a source of future embarrassment to BNFL. Given the political turmoil in the Middle East and its history of armed conflict, Capenhurst's by-products may one day be blowing up the tank crews of Israel's enemies.

Non-nuclear weapon states

In addition to the five nuclear weapon states – The United States, the Soviet Union, Britain, France and China – several others, including Israel, South Africa and India are assumed to have nuclear warheads or the capability of making them. Other countries – notably Pakistan and Iraq – are known to be trying or have declared an interest in developing them.

It is frequently argued that the spread of nuclear technology is the greatest aid to the proliferation of nuclear weapons: if a country can build an enrichment or reprocessing plant, it can produce its own fissile materials. It still needs, however, the radioactive raw materials. As a result, the trade in yellowcake and fuel materials like hex must also be seen as sensitive: materials supplied for civil purposes could end up in a military programme. In theory this should never occur. If countries do not already have nuclear weapons, the transfer of nuclear materials from civil to military use should be prevented by the Non-Proliferation Treaty (NPT). Exporting countries also usually require peaceful assurances as a matter of policy. This is supposed to be the case in Britain where: "It is . . . the policy of Her Majesty's Government to permit the export of power reactors and fuel only when there are adequate safeguards to prevent the use for a military purpose of the fissile material obtained from them".[45]

Apply this to the Superphénix where British-supplied plutonium could help France to produce weapons material and the limitations are obvious. (In any case, as an existing nuclear weapon state, France, like Britain, is entitled under the Non-Proliferation Treaty to withdraw nuclear materials from safeguards.) Government assurances that materials supplied to France will be used only for peaceful purposes[46] are not entirely convincing.

Furthermore, not all countries have signed the NPT. Absent signatories include Argentina, Brazil, Chile, India, Israel and Pakistan, many of which have aspired to acquire nuclear weapons. Failure to sign is not necessarily a deterrent to trading with Britain's nuclear industry. In 1982, for example, the UKAEA signed a contract to supply research reactor fuel to the Chilean Atomic Energy Authority. Deliveries of the enriched uranium were completed in 1984.

Although Chile has not been viewed as a potential nuclear power – at the time merely a military dictatorship – other non-signatories have. Brazil, when it too was under military rule, acquired most of the technology for nuclear self-sufficiency – reactors, enrichment and reprocessing plants; largely from a deal with West Germany. With its own reserves of uranium, it also had an avowed interest in peaceful nuclear explosions – which can of course be used for non-peaceful purposes – and ambitions to build nuclear powered submarines. Despite the military overtones of the country's nuclear industry, the British, Dutch and German consortium Urenco signed a contract in 1977 to supply low enriched uranium to Brazil. The military implications were recognised at the time by the Dutch parliament, which demanded extra safeguards to prevent the material being used by Brazil for military purposes – assurances which Britain and West Germany regarded as unnecessary. The Brazilian deal became one of the issues in a "Stop Urenco" campaign in the late 1970s. Tactics included a one-day vehicle blockade of Capenhurst by protestors to stop traffic entering and leaving the site. Dutch reservations were eventually sidestepped by Urenco's announcement in 1981 that deliveries would be made from Capenhurst, rather than the Dutch plant at Almelo.

And so in the following year, the first deliveries of uranium hexafluoride, enriched to about 3.5 per cent U235, were shipped from Capenhurst to Brazil.[47] The contract for 3,901 tSWU of enrichment (of Brazilian uranium) ran until 1990. Although the material is officially destined for Brazil's fledgling nuclear power programme, Dutch fears that Urenco's deliveries could eventually produce weapons-grade materials remained as valid as ever. With its own enrichment and reprocessing facilities, Brazil has had two potential routes to nuclear weapons if it had been prepared to ignore or cheat on safeguards. Either one could utilise, directly or indirectly, the material arriving from Britain: irradiated fuel made from Urenco's enriched uranium could be reprocessed to obtain plutonium, or the hex could be fed directly into the enrichment plant to produce highly enriched uranium. In the event, uranium enrichment has not excceeded 20 per cent and the country's military government was replaced by the civilian administration of President Fernando Collor de Mello who pledged Brazil's nuclear industry to peaceful use.

Unfortunately, other countries continue to pursue military nuclear ambitions and these require nuclear materials as well as the processing technology. Iraq is the best known example. During the Gulf War its small stock of highly enriched uranium (obtained from France and the Soviet Union for research reactors) received much attention. Although enough for at most one warhead, Iraq's ambition to develop nuclear weapons was clear, as IAEA inspections subsequently revealed.

Yet while the seizure of warhead "trigger" components at Heathrow Airport made headlines, Iraq's acquisition of nuclear fuel materials had gone largely unnoticed. Over the years, Iraq had obtained supplies of uranium ore from Portugal[48] and Niger.[49] Three hundred tonnes of yellowcake came from Brazil under a secret bilateral agreement between the two countries officially unacknowledged until 1990.[50] Iraq had also acquired low-enriched uranium recovered by reprocessing enriched uranium fuel. Such materials – some of which had also been supplied from Italy and West Germany[51] – are the requirements of a country with nuclear power stations which Iraq has never been. They would nevertheless have come in useful for military purposes if its enrichment plant had ever been built. Also useful might have been certain exports from Britain, namely "less than 100 kgs" of depleted uranium (as used in plutonium breeder fuel)[52] and a small quantity of plutonium-238 from Amersham International, potentially useful for callibrating radiation monitoring equipment.

The general danger is that when nuclear materials are transported to one country that may not be the end of the journey. Brazil has had links with Argentina, another South American country whose nuclear ambitions have in the past been viewed with concern. Brazil has had links with Iraq; Iraq has links with Pakistan – a country which *has* succeeded in building an enrichment plant and is regarded as a potential nuclear weapons state. Pakistan has links with Libya (from where it has obtained uranium), and so the connections go on.

This world of secret trade deals, covert weapons programmes and bilateral agreements between nations with military ambitions, raises justifiable fears about nuclear trade in general. Can safeguards ever be trusted when the countries involved are paranoid about national security? Even in Britain, the military nuclear programme casts doubt on the use of certain ostensibly civil nuclear materials: the restrictions on information about plutonium accountancy, for example, make it impossible to independently verify government and industry reassurances of their peaceful use. Is any other government with a nuclear weapons programme likely to be more open? The general problem is that safeguards assume trust – which can never be guaranteed when non-nuclear weapon states pursue nuclear weapons programmes.

The transport of materials like yellowcake and hex may not make the only contribution to a country's military effort but they could be the vital ones. For anyone worried about the capacity of the human race to destroy itself, no effort will be spared to prevent that disaster. Whatever contribution the nuclear industry makes to the weapons programme, in Britain or elsewhere – be it 10 per cent or just one per cent of business – if opposing the entire industry helps to reduce that threat, then opposing the entire industry will be justified.

13   Radiation

Underlying public concern about all things nuclear are issues of health and safety; and underlying those are the hazards of radiation. In practice, the amount of radioactivity in many consignments of radioactive material – radioisotopes in particular – is relatively small, if not minuscule. By contrast, other shipments, for example spent fuel, contain millions of becquerels of radioactivity. Hence the concern that radioactive materials in transit might be a health hazard for people close to the transport routes.

Such fears were voiced, for example, in 1986 when the *Daily Mirror* highlighted clusters of leukaemia cases in the Somerset district of Woodspring. The newspaper observed: "There is a strange link between the four leukaemia clusters in Woodspring. All lie close to the railway line on which irradiated waste (spent nuclear fuel) is transported twice a week from Hinkley Point to Sellafield."[1] Similar fears of an increased incidence of leukaemia have been voiced along other flask routes, including the lines through North London.

If such fears have any foundation, how could a health hazard arise? Do radioactive materials leak en route? In a sense many do, although not necessarily as a result of an accident: even under normal transport circumstances, some containers of radioactive material give off small doses of gamma radiation and/or neutrons. (Alpha and beta particles will not penetrate transport containers.) Despite the emphasis on structural strength and safety, transport containers are *not* required to be radiation-free on the outside. Instead, the regulations set limits which define maximum permissible emissions in normal transport circumstances and accidents. Thus even without an accident, the thickest transport containers in use – for example, a 14-inch thick spent fuel flask – still allow a tiny proportion of radiation to escape.

Allegations of any health hazard are dismissed by the industry itself. Radiation doses from the routine transport of radioactive materials were assessed by the NRPB in 1982. The conclusions of their report, *Radiation exposure resulting from the normal transport of radioactive materials within the United Kingdom*, showed that doses from transport are far less than the doses received by the public from other sections of the nuclear industry including power stations and the discharges from Sellafield.[2]

In fact, the largest source of radiation exposure for the public from the transport of radioactive materials is from the transport of radioisotopes. Figures given in the NRPB report – the only comprehensive assessment to date – enable a comparison to be made of the doses from different types of radioactive material. The maximum individual dose to a member of the public from the movement of radioisotopes was estimated at about 40μSv per year. A more recent assessment of one mode of transport – air travel – has identified higher dose situations. This study, published in 1990, calculated that regular airline passengers – defined as those who made 40 flights a year – on planes regularly carrying radioactive materials, could, depending where they sit, receive individual maximum doses of 100 μSv per year.[3]

By comparison, maximum individual doses from spent magnox fuel were estimated at only 2μSv a year. Moreover, such a dose, according to the Board, is only likely to be received by people living near to locations like marshalling yards where flasks are stationary. Train timetables also provide "layovers' within the journey during which flasks are parked in sidings, sometimes for almost half a day. Several locations are regularly used which provide more opportunities for longer radiation exposures. Elsewhere, most individual doses from the normal transport of spent fuel should be far lower.

To get an idea of the risks associated with the maximum doses from transport, the figures above – measured in *micro*sieverts – can compared with the NRPB's recommended dose limit for members of the public. This is currently 1 *milli*sievert per year, although the Board advise that exposure from a single site should not exceed 0.5 mSv per year. (This is to allow for the fact that members of the public who receive relatively high doses from a single site may also receive radiation from other sources; the 0.5 mSv limit is intended to ensure that the total dose is still below 1 mSv.) According to the NRPB, the advisory limit of 0.5 mSv constitutes a one in 70,000 risk of dying of cancer. Therefore, the maximum individual dose received by certain well-travelled airline passengers – 100 μSv – constitutes 20 per cent of the advisory limit or a one in 350,000 chance of dying of cancer. A 2μSv dose from a spent fuel flask is only 0.4 per cent of the limit – or a one in 17,500,000 chance of fatal cancer.

On the face of it, the NRPB's confidence would appear to be justified. Nevertheless, while the contribution from materials like spent fuel may not be "significant" to problems such as leukaemia, there is a remote possibility of a connection.

Occupational doses

The situation is not always so reassuring for those workers involved in the transport of radioactive material. Estimates of what are called "occupational" doses were included in the NRPB's 1984 report on the normal transport of radioactive materials. According to the Board, the annual collective occupational dose was determined to be about one man-Sv. The major source of exposure – 87 per cent – is from the movement of radioisotopes. The remainder came from the transport operations of the nuclear fuel cycle, principally gamma emissions from uranium products.

Almost half the collective occupational dose came from the road transport of just one item – technetium generators – which contain the gamma emitting isotope molybdenum-99. Radioisotopes also account for the highest individual doses. Road transportation inevitably means that vehicle drivers are in relatively close proximity to their radioactive loads and long journeys result in correspondingly long exposures: for some isotopes, dose rates in drivers' cabs may be up to 80 μSv per hour. According to the NRPB: "The high mileage and the nature of road transport operations result in a very small number of individual occupational doses exceeding 15 mSv a year".[4] Fifteen mSv per year is the NRPB's recommended maximum average annual dose for nuclear industry workers and constitutes a fatal cancer risk of one in 2,000 for continuous exposure.

Doses are also received during loading and unloading operations where workers are physically handling radioactive packages or come close to them with the help of loading equipment. With surface dose rates from some radioisotope packages reaching 1.5 mSv per hour, these operations can lead to accumulating exposures.

Of the contribution made by the nuclear fuel cycle, over half came from the transport of spent fuel. The rest were from shipments of uranium ore imports, uranium hexafluoride and oxide, new fuel elements and radioactive wastes.

For spent fuel flasks, the highest exposures occur at the railheads where flasks from British nuclear power stations are loaded onto railway wagons prior to being dispatched. With the help of a travelling crane, flask loading is undertaken by "slingers" whose job is to manouevre the flask, suspended from a crane, into position on its flatrol. Slingers probably receive the highest regular doses, although according to the NRPB, this is unlikely to exceed 100 μSv per year.

Railway workers at marshalling yards may also receive small doses of radiation. Most flask wagons must be routinely coupled and uncoupled on their journey to Sellafield. Flatrols are marshalled together along the route, engines must be changed, and guards vans and barrier wagons are added or removed – operations which bring railway workers into close proximity to flasks. Minor repairs may also be necessary to flask wagons which could further expose BR staff to radiation.

Overall, however, the published evidence indicates that the normal transport of radioactive materials gives lower doses of radiation to both workers and the public than many other sections of the nuclear industry. While transport and nuclear industry workers are at more risk than the general public, official figures suggest that neither group should have too much to fear. However, such statistics do not reassure everyone. The counter argument is that even the smallest doses of radiation can be potentially harmful and, in the interests of the best possible good health, they must be opposed as a matter of principle. There are two other reasons for concern. First, official statistics are incomplete and do not tell the whole story. Second, worries about radiation exposure from transport operations have been influenced by the wider controversy over the hazards of low-level radiation.

Routine hazards

Official reassurances about radiation exposure are undermined by nuclear secrecy: for some radioactive materials, radiation doses are secret. In particular, there is no published record of emissions from the transport of military nuclear materials. Although information from the MoD was incorporated in the NRPB's overall assessment, no details were included in the published results. Despite the comprehensive title of the NRPB report – "Radiation Exposure Resulting from the Normal Transport of Radioactive Material within the UK" – there was no reference to new or used fuel from nuclear submarines, to highly enriched uranium or plutonium in any shape or form, or to nuclear warheads themselves.

In practice, military materials move around in smaller quantities, and less frequently than the tonnages required for civil nuclear power. Moreover, many are predominently alpha emitters, giving off little or no radiation that would penetrate a transport container. Nevertheless, their omission means that materials are moving around whose contribution to collective public exposure is unknown – at least, to the general public.

Even regarding civil materials, the official picture is incomplete because the normal transport of radioactive materials is not always normal: the public's collective dose can be increased by radioactive contamination on the outside of transport containers. This arises not from any transport accident but is typically picked up during the loading and unloading of transport containers.

External contamination has been associated in particular with spent fuel flasks. These are usually loaded under water, an operation which involves dunking the entire flask into the cooling ponds where irradiated fuel is stored. As the pond waters are also radioactive, wet flasks must be decontaminated and dried after loading, an operation which is not always totally successful. In some cases, traces of radioactivity remain trapped in crevices on the exterior surface. A more intriguing phenomenon has been called "sweating out" – radioactivity from cooling ponds is absorbed into the paintwork on the outside surface of a flask and slowly released during the subsequent journey. In this way, the outside of a flask which is dispatched supposedly clean becomes more radioactive en route; a problem exacerbated, according to the UKAEA, by humid weather and rain.[5]

These problems have persisted for many years. In 1980, for example, NTL reported that "sweating out of contamination continues to cause difficulties . . .".[6] In the same year, the UKAEA noted that surface contamination on the Winfrith SGHWR flask increased during journeys "on many occasions", with external radiation levels sometimes rising above the IAEA's regulatory limits for surface contamination.[7] Although relatively minor, such problems are not supposed to happen at all and in that sense, they might be regarded as "accidents".

Most contamination has been associated with flasks belonging to the former CEGB. By 1983, the problem had become sufficiently serious to justify a secret BNFL and CEGB joint "Panel of Investigation" chaired by Mr J.K. Donoghue, BNFL's head of Safety and Medical Services at Sellafield. The resulting confidential report was leaked to the press in 1984. The report confirmed many previous suspicions, although its assessment of the hazard was reassuring: "The magnitude and scale of these contamination events is such that they present insignificant radiological risk to those personnel properly involved with handling the flask traffic and to the general public".

However, other parts of the report gave cause for concern. Although the number of cases of contamination was apparently decreasing, the statistics gave, for the first time, an indication of how extensive the problem had been. Between 1980 and 1982, the number of flasks exceeding the "derived limit" (from national and international regulations) fell from 20 to 10 for flasks received at Sellafield and from 40 to 20 for flasks received at nuclear power stations. The figures identified a total of 90 cases of contamination, although the levels of radioactivity found were not recorded.

The report contained a further surprise: problems were not confined to loaded flasks of fuel. As the figures above show, twice as many contaminated flasks were detected on arrival at the power stations – flasks which would have been empty. Thus contrary to what might be expected, an empty flask is more likely to be contaminated on the outside than a full one. Given Sellafield's reputation for rampant radioactivity, such problems might not seem surprising although the report was inconclusive about the cause.

Nor could it explain another phenomenon that might worry railway workers: contamination had also been detected on the railway wagons which transported the flasks. Despite the fact that flatrols have no direct contact with primary sources of contamination (such as cooling pond water or skips), flatrol couplings, hoses and brakewheels – the parts handled by railway personnel – had all on occasions been contaminated. As with the flasks themselves, there were more cases of flatrol contamination on arrival at the power stations than at Sellafield.

Is such contamination a hazard? Although the Donoghue report was never officially published, the NRPB have estimated that the release of contamination during flask journeys – giving a so-called "cloud dose" – accounts for over 97 per cent of the public's annual collective dose of 19.5 man Sv 10-^{3} from the transport of irradiated magnox fuel; the remaining dose is given by direct gamma emissions from flasks. Moreover, 84 per cent of the cloud dose arises when flasks are stationary.[8]

However, as the NRPB admit, there is no measured data against which such estimates can be checked – the reality may be better or worse. Several reports of environmental contamination in the past have been blamed on "leaks" from transport flasks. In 1981, for example, the CEGB excavated a section of roadway at the Trawsfynydd nuclear power station which had been contaminated by caesium-137, a radioactive substance found in cooling ponds which could have dripped from a spent fuel flask.[9] In 1989, traces of caesium-137 were found when Strathclyde Regional Council analysed samples of railway track ballast from Fairlie station in Scotland. The station, which has a regular passenger service to Glasgow, is also the railhead for spent fuel flasks from nearby Hunterston power station.[10] Another example returns us to the alleged links with leukaemia cases in Somerset. In 1988, the CEGB removed ballast contaminated with caesium-137 from the railway sidings at Bridgwater; not only the location where spent fuel flasks from Hinkley Point arrive and depart for Sellafield, but close to a local school.[11]

Low-level radiation

Some years ago an unusual medical statistic was noticed in Cumbria: "Among the 2,000 odd inhabitants of Seascale, near Windscale, it is reported that there were 4 child leukaemia deaths . . .". According to the author, the occurrence of four deaths in the same place in the same year amounted to a leukaemia death rate 36 times the national average. The suggested reason for the fatalities can be guessed from Seascale's location: the proximity to Windscale was not necessarily a coincidence: "It is also reported that the sandy beaches in the neighbourhood are well known to be slightly contaminated by active waste which is pumped well out to sea".[12]

Sounds familiar? In fact, those observations were made in a book published in 1958 and the statistics quoted were for 1956. Three decades later and the problem was rediscovered by Yorkshire Television for its programme "Windscale – the Nuclear Laundry". The television company identified many more cases of leukaemia, not only in Seascale itself, but also in the surrounding areas. Similar health problems have also been alleged near other nuclear installations, including Dounreay, Sizewell and Hinkley Point, and near the MoD's warhead factories at Aldermaston and Burghfield.

Yet despite the accumulating evidence, the nuclear industry continues to deny responsibility. No one disputes that massive doses of radiation can be deadly: exposure to hundreds of sieverts within a few minutes will kill in a matter of weeks. However, cancers arising from small doses of radiation can take decades to appear. This long latency period, which helps disconnect cause from effect, ensures that the cancerous consequences of the nuclear industry get lost in the mists of time. However many eventually die from radiation-induced cancers, their suffering will be hidden among millions of other victims buried by cancers from other causes. Moreover, there is no indisputable way of linking an individual cancer to a particular cause. Radiation exposure is not the only cause of cancer and causal evidence – however strong – will ultimately be circumstantial. Thus the nuclear industry has been reasonably confident that if its operations are harmful, it cannot be proven guilty.

The growing evidence of health problems parallels a changing attitude to the hazards of low-level radiation: many experts now believe that it is more dangerous than previously assumed. In the early years of the industry, a low dose of radiation was often considered harmless: in effect, there was a dose "threshold" below which exposure would produce no harm. This was the view, for example, of the 1958 Fleck Report, set up by the government of the day to examine health and safety controls within the UKAEA. The Fleck Report noted that the standards set by the International Commission on Radiological Protection (ICRP) ". . . prescribe the *maximum safe dose* for an individual and from this it is necessary to derive codes of practice . . ." [author's emphasis].[13]

That clear assumption from the fifties – that there is a "maximum safe dose" of radiation – can be compared to a more recent view from the NRPB: "At present, there is no sound basis for assuming that there is any threshold of dose below which neither cancer nor hereditary defects can be caused by radiation." Therefore, the NRPB conclude: ". . . any dose, no matter how small, creates a finite risk of cancer or hereditary defects."[14]

Individual radiation doses from the normal transport of radioactive materials are sometimes compared with doses from natural radiation, which in Britain average 2.2 millisieverts per year – doses from transport are much lower. However, attitudes towards the risks of natural radiation have also changed. In 1974 the UKAEA's official historian, Margaret Gowing wrote that although countries like India and Brazil had higher levels of background radiation than Britain: "Investigations have not shown any correlation between levels of natural radioactivity and the incidence of genetic defects or diseases associated with radiation".[15] Two years later, the Flowers Report – the government Royal Commission on Environmental Pollution – also referred to background radiation in India, but for a sadly different reason: "A recent paper has shown that there has been an increase in severe mental retardation and in chromosome aberrations in children born to mothers in Kerala, South India, where background radiation doses are between 1.5 and 3 rem/year (0.015 and 0.03 Sv/year). The incidence is particularly high among children born to mothers over 30, who would have received higher doses."[16]

Radiation and pregnancy don't mix! A container of yellowcake parked on the outskirts of Preston, presumably on its way to Springfields. With a transport index of 6, the radiation dose at 1m from the container would have been 0.06mSv per hour

None of which is to suggest that a small dose of radiation is automatically harmful. The vast majority of people irradiated even as a result of the accidents at Windscale or Chernobyl will not die of cancer. There is also a well-entrenched view that cell damage from small doses can be repaired by the body. Scientific opinion has nevertheless changed over the years as knowledge about radiation has increased – what was once considered safe is now known to carry a risk, a trend reflected in the dramatic reductions over the years in the ICRP's own guidelines on radiation exposure. Until 1950, for example, the ICRP was recommending a maximum permissible dose for radiation workers of 52 roentgens (45.76 rems) per year.

For members of the British public, the ICRP's current recommended annual dose limit (from sources other than medical and natural background radiation) was lowered in 1985 from 5 mSv to just 1 mSv. Two years later a further reassessment was made. After retrospectively revaluating data on the Hiroshima and Nagasaki survivors, the ICRP announced that the estimated risk for the exposed population should be doubled. In other words, all previous calculations of the relationship between dose and risk, derived from the Japanese experience, had been underestimated by 50 per cent. Although the Commission considered that this did not justify a change in their recommended dose limits, the NRPB calculated that the risk of cancer associated with 1 mSv per year should be increased by a factor of three from 1 in 100,000 to 1 in 33,000.[17] In other words, exposure to radiation was now considered three times as dangerous as before. As far as the public was concerned the 1 mSv limit was still considered adequate, but a lower limit was recommended for the most highly exposed groups (those exposed to effluent discharges) of 0.5 mSv per year from any single site. This allowed for the fact that such people might also receive radiation from other sources thereby increasing their total dose above 1 mSv.

These downward adjustments are hardly reassuring. After all, had a member of the public received, say, 2.5 mSv in 1984, he or she would have been told that they had received only 50 per cent of the recommended maximum permissible annual dose – the implication being that there was little to worry about – yet one year later they would have been two and a half times over the limit! The reduction to 1 mSv has created just such an anomaly with regard to transport. At the time of writing (1991), radiation emission limits for category 3 yellow packages – which include, for example, spent fuel flasks – still allow a dose limit of up to 2 mSv per hour at their surface. This maximum level, set by the IAEA before the 1985 reduction, is now twice the annual exposure limit recommended for the British public by the NRPB. These trends introduce an element of uncertainty regarding current official attitudes towards the risks of radiation: as scientific opinion on the subject has changed in the past, might it not change in the future? Will radiation doses now officially condoned eventually be shown as unacceptable?

Research by independent experts – not necessarily critical of the nuclear industry *per se* – has suggested a need for even more reductions to dose limits. For example, the Sizewell Inquiry heard evidence from Dr Alice Stewart that both the ICRP and the NRPB had been underestimating the effects of low level radiation by a factor of between 10 and 20.[18] More recently, the MRC published a study of deaths among 39,546 people who had worked for the UKAEA up to 1979.[19] The study found an excess cancer mortality rate (of 12.5 deaths per million person-years per rem for all cancers) intepreted by one independent expert as between five and six times that implied by current ICRP recommendations.[20] In 1990, a study by Professor Gardner from Southampton University suggested that the accumulation of radiation exposures of 100mSv or more by Sellafield workers prior to becoming parents, was statistically associated with a six- to eight-fold increase in the risk of leukaemia in the offspring.[21]

These and other studies suggest that low level radiation will continue to be seen as potentially more dangerous than previously believed. This has clear implications for those concerned about emissions from radioactive materials in transit: the apparently small radiation doses received from moving sources – small, that is, relative to other emissions from the nuclear industry – may turn out to be more hazardous than currently assumed.

14   Accidents

The possibility that an accident might release substantial quantities of radioactivity has been the major concern about the transport of radioactive materials. Past experience shows that accidents do indeed happen, and some have resulted in radioactive leaks or spillages. Fortunately, most which have occurred to date have involved consignments of relatively low level radioactive materials, notably radioisotopes and yellowcake.

Radioisotopes, for example, have been involved in accidents on land, in the air and at sea. Road accidents have ranged from the trivial – minor bumps causing only superficial damage to the packaging – to the serious, with transport vehicles overturning or being burnt out by fire. According to Amersham International, one of the worst involving their products occurred in 1978. A vehicle delivering four caesium-137 sources was "completely destroyed" in a road accident and the driver killed. Fortunately, although the packages containing the caesium were "severely dented" there was no loss of radioactivity.[1]

One of the worst air accidents affecting radioisotopes occurred in the following year. On 7 October, a Swissair DC-8 skidded off a wet runway at Athens and crashed onto an adjoining public road. Fourteen passengers were killed when the aircraft burst into flames. Recovered from the burnt-out wreckage was part of a consignment of plutonium isotopes prepared by Britain's Radiochemical Centre (now Amersham International) and bound for medical use in China. "Part of" because the rest of the plutonium – eight pellets of Pu-238, with a total radioactivity of 12 GBq – along with lead from the containers in which they had been packed, was irretrievably lost in the debris.[2] Incidents such as the sinking in 1988 of the *Ardlough* on its way from Liverpool to Belfast with the loss of a consignment of californium-252 (a radioisotope used by the Ministry of Defence) show that the marine environment is no less fraught with danger.[3]

In addition, incidents can occur which are not transport accidents as such but accidents which occur in transit. The distinction is illustrated by a Chinese incident: over 60 miles of road were contaminated when luminescent powder containing radium-228 leaked from a broken glass ampoule being transported by lorry. Clearly, the effects of a leak – potentially dangerous even if the source is stationary – will be far more widespread when the source is on the move. Although the contamination on that occasion was not serious (according to the Chinese nuclear industry), transport accidents have reportedly "had a considerable influence on public psychology" in that country.[4]

The Chinese incident shows how dangers can be exacerbated when problems are not noticed or ignored. A British incident in 1974 provides another example. A vehicle crushed a package containing the radioisotope yttrium-90 while being loaded onto a cargo plane at Gatwick Airport. Despite being damaged, the package was flown to West Germany. On landing at Dusseldorf, the crushed package – which should never have been loaded in that state – was found to have leaked radioactive liquid onto the aircraft floor. The plane was forced to return to England for decontamination and the replacement of part of its floor.[5]

At least on that occasion action was taken at the end of the affected journey. In the same year in the United States, a combination of inadequate packing and improper stowage resulted in 250 cm^3 of a solution of molybdenum-99 leaking from a Type B package on a passenger flight from New York to Houston. Not only had the gamma-emitting liquid been transported in a polythene bottle with a leaky lid, but this in turn had been loaded inside a steel and lead container which had part of a gasket missing between its body and lid. With the container placed – wrongly – on its side during the flight, the combination of three separate faults resulted in radioactive liquid leaking into the cargo hold of the plane. On this occasion, however, the contamination went unnoticed and after the container had been unloaded, the plane had made nine more flights before the radioactivity was detected. As a result of this oversight, more than 900 passengers and their luggage were carried on a radioactive plane and exposed to radiation even after the source itself had been removed.[6]

In the case of yellowcake, the large number of consignments moving around, especially in uranium mining areas, means that these too have had their share of misfortune. A 1985 crash in the United States demonstrates what can go wrong. Truck driver Beryl Kaplar was killed when her vehicle, carrying 53 drums of Canadian yellowcake, smashed into the side of a train on a North Dakota level crossing. The uranium ore had been on its way from Saskatchewan to the Sequoyah Fuels conversion plant at Gore, Oklahoma. The impact, severe enough to derail the locomotives, ruptured 30 of the drums and spilled some 20 tonnes of yellowcake. In a clean-up operation lasting several days, yellowcake had to be removed from the locomotives by sandblasting. The rest was recovered by vacuuming.[7]

In 1979, BNFL revealed a history of spillages affecting the transport of yellowcake in Britain. In the company's words: "For twenty years yellowcake was delivered to Springfields as loose loads of 45-gallon drums. As a consequence two or three times a year a drum spillage incident occurred when a drum fell into a dock or onto a public roadway."[8] In one such incident in Preston, five drums of yellowcake broke open after a traffic accident and the road had to be decontaminated.[9] However, since the containerisation of uranium imports in 1978, only one transport incident involving yellowcake has been acknowledged. (A freight container and drum were punctured by a forklift truck.)[10]

Fortunately, yellowcake, like most radioisotopes, is a relatively low level radioactive hazard. There have been far fewer accidents affecting the more dangerous radioactive substances such as spent fuel. Moreover, where accidents have occurred, most have been relatively minor events with no releases of radioactivity. Perhaps, it might be said, because more care is taken with their transport and they are carried in more robust containers. Yet even consignments of spent fuel have been involved in accidents resulting in radioactive releases. Despite the assertion by Britain's nuclear industry that flasks of spent fuel have never leaked, the UKAEA itself has in the past reported two instances of accidents (location unknown) resulting in flasks leaking contaminated cooling water.[11]

An indication of the nature and number of accidents in Britain has been given by the National Radiological Protection Board (NRPB) which has reviewed the accident record from 1964 to 1989.[12] Although the NRPB ackowledge that their data is incomplete and advise caution in its use, their reports provide the most comprehensive published assessment of the accident record in Britain: out of an estimated 900,000 or so transport movements during the period reviewed, the NRPB identified 388 "events" – the collective term for accidents and incidents during transport, including loading and unloading operations. (The NRPB define accidents as involving "significant damage to the transport or package" while incidents refer to other transport events "for example, theft, incorrect procedures, delays and incorrect packaging prior to shipment". Table 14.1 gives a breakdown of the materials involved.

The figures show the prevalence of accidents involving radioisotopes and radiographic sources, which account for two-thirds of all reported events. Up to 1983, almost half of all events – 140 reports – involved radioisotopes in the cargo area at Heathrow airport. Most occurred when packages fell from pallets and were run over by fork-lift trucks or other vehicles. Only in one instance was radioactivity released. Loaded nuclear fuel flasks account for 39 incidents, although most involved low speed collisions or derailments, principally at marshalling yards. (Events involving empty flasks were classified under "Residues".)

Clearly, the number of accidents to date has been extremely small in relation to the far larger number of shipments completed without incident.

It must also be recognised that containers of radioactive materials can be involved in major transport accidents and *not* release radioactivity. Indeed, some of the more spectacular accidents in the past – especially in the United States - provide evidence that packages of radioactive material can survive serious accidents. In 1970, for example, a lorry carrying a spent fuel flask near Oak Ridge, Tennessee, overturned and crashed after the driver swerved to avoid a car. The driver was killed and the flask thrown 100 feet before landing upside-down in a ditch.[13] In the same year in Connecticut, the left rear wheels came off a truck carrying drums of highly enriched uranium-zirconium alloy scrap (the sort of material that would originate in the nuclear submarine programme). The truck and its cargo veered off the road, again landing upside-down in a ditch.[14] In North Carolina, a 1977 train crash resulted in the derailment of several wagons and their loads including four cylinders of unenriched uranium hexafluoride. Reportedly, a fire broke out at the scene of the accident.[15] In 1968, a Boeing 707 carrying a Type A package of radioisotopes, burst into flames after landing at Heathrow Airport. Five people died as the plane was destroyed by fire.[16]

Table 14.1: Transport accidents and incidents, 1964–1989

| Type of material | Transport | | Handling | | |
	Accidents	Incidents	Accidents	Incidents	Total
Yellowcake	2	7	12	1	22
Pre-fuel material	5	3	0	2	10
New fuel	1	3	0	0	4
Irradiated fuel	3	29	0	7	39
Residues	4	30	2	6	42
Radioactive wastes	3	5	0	2	10
Radioisotopes	19	19	147	31	216
Radiography sources	8	18	2	17	45
Totals	45	114	163	66	388

Source: NRPB

Yet in all these major accidents, the respective containers and packages, although dented, damaged or charred, survived sufficiently intact to prevent their radioactive contents escaping. The accident record to date might therefore be viewed with ambivalence – either reassuring or a source of concern. On the one hand, some of the worst accidents of the past support the industry view that the transport of radioactive material is basically safe – safety is secured by the integrity of the transport packaging rather than the transport environment. On the other hand, the record shows that leaks, spillages and abnormal radiation doses can and do indeed happen despite efforts to ensure that they do not.

More important perhaps is whether past experience determines the future. Accidents are by their very nature unplanned and unpredictable and there can be no guarantee that a serious accident in the future will not release substantial quantities of radioactive material. If there is any possibility of that occurring, the implications for public health will inevitably be a matter of public concern.

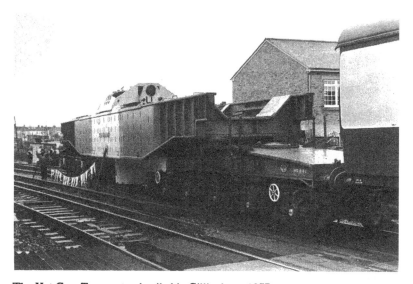

The Hot Core Transporter derailed in Gillingham, 1977

The consequences of an accident

What might the consequences of a major accident be? With relatively few serious accidents to learn from, past experience gives little guidance. In Britain, the only assessment of the radiological impact of transport "events" to date is the NRPB review. Of the 388 accidents and incidents recorded, only 65 were classified as "significant", defined as one that led to "an increase in dose rates above that normally expected from the transport of the type of package". The total collective dose to workers from these 65 events over twenty five years was calculated as 6.7 man Sv. For members of the public, the collective dose was just 0.026 man Sv.

Significant events were mostly spillages of radioactive material such as yellowcake, or incidents, typically involving radiography sources (which emit gamma radiation) where higher than normal radiation doses resulted from exposing the source but without necessarily releasing radioactive material. Referring to spillages of yellowcake, the Board describe the consequences as:

> ... light contamination, low internal and external exposure of drivers and members of the public resulting in low radiological consequences, and with no exposure exceeding 2 mSv for dock workers or drivers and 0.01 mSv for a member of the public.

Higher doses have been caused by radioisotopes and especially radiography sources. Indeed, according to the NRPB report, for both the public and the workers concerned, events involving radiography sources – of which just 17 were classified as significant – accounted for over 99 per cent of the total collective dose received during the review period. Other radioisotopes accounted for the remaining one per cent.

Significant events involving radiography sources included instances where radioactive materials such as iridium-192, caesium-137 and cobalt-60 had been transported without being properly shielded. The highest radiation doses would have been received by people in the vehicles carrying the sources – typically just a few feet away. The worst case recorded by the NRPB concerned a 190 GBq iridium-192 site radiography source which fell out of its container and remained exposed for 17 hours before the error was discovered. As a result, three workers received whole-body doses of 52, 510 and 2,380 mSv – the last over 150 times the maximum annual limit recommended for nuclear industry workers in Britain.[17]

Such incidents can also affect the public. A similar problem occurred in the United States, when a used iridium-192 radiographic source was loaded the wrong way round into its container. With the radioactive source almost completely exposed, the container was flown back to its manufacturer on a US passenger flight. Subsequent investigations found that radiation doses received by passengers during the flight could have been as high as 0.13 Sv – that is, 130 times the recommended annual radiation limit for the British public.[18]

The NRPB's assessment indicates that nothing comparable has yet affected the public in Britain. The highest doses have arisen from the same transport events which have irradiated workers, namely problems with radiography sources. On several occasions, exposed sources have been left in vehicles parked overnight near houses, thus giving a continuous dose of radiation to nearby residents. However, although the NRPB acknowledge that members of the public have been irradiated by unshielded sources, "none . . . have exceeded a radiation dose of 2mSv from any single event".

2mSv is, nevertheless, twice the present annual radiation limit for the British public – a limit recommended by the NRPB itself. Regarding transport accidents and incidents overall, the NRPB's conclusion is reassuring: "collective doses are low compared to exposures from the normal transport of radioactive materials and are extremely low compared to exposures from natural sources of radiation".

Outside Britain, most accidents resulting in radiation exposures to the public have also involved materials such as yellowcake and radioisotopes. Thus when it comes to estimating the consequences of a substantial release of more dangerous radioactive materials – for example, plutonium or spent fuel – there is little empirical information to go on. Predictions of what might happen in more extreme circumstances have been plagued with uncertainty, disagreement and a welter of conflicting opinions on all aspects of the subject: the types of accidents that might occur; the likelihood of such events; and the radiological consequences.

The industry contends that severe accidents would cause nothing worse than minor problems of leakage with a minimal release of radioactivity. Radiation exposure would be negligible, if that, and certainly nothing to worry about. No other scenario is allowed.

Assessments of the consequences of a leak from a spent fuel flask show how the industry and its critics can reach diametrically opposed conclusions. For example, computer predictions by the NRPB illustrate the industry's optimism. One was carried out in 1983 at the request of the Greater London Council (GLC). In connection with the Sizewell B Inquiry, the GLC commissioned the Board to assess the consequences of a leak in London from a flask of PWR spent fuel from the proposed reactor. The NRPB study made use of a computer programme known by the acronym MARC – Methodology for Assessing Radiological Consequences. A number of accident scenarios were considered, including a severe impact followed by a two-hour 1000°C fire affecting a stationary flask in the marshalling yards at Willesden Junction. The assumed result of the accident was a tiny release of radioactivity. Although this included a proportion of the many different substances present in spent fuel (specified as noble gases, caesium/ruthenium, all other fission products, actinides and cobalt), for each type of material the amount released was well under one percent of the total inventory. The predicted health effects were a correspondingly tiny increase in the public's risk of cancer, summed up by the NRPB as "of no great radiological concern". Risks were presented as a range of probabilities, the worst of which would be equivalent to 18 fatal cancers in a population of 55 million. In relation to west London the hypothetical death toll would effectively be zero.[19]

On other occasions, industry reassurances are taken to extremes. A CEGB public relations leaflet on spent fuel transport provides a classic example.

Assessing the likelihood of an accident, the question-and-answer format includes the following:

Q. Even if the chances of an accident are so remote what happens if there is a crash and a flask cracks or leaks?

A. No one would come to any harm if a flask cracked. If the water in the flask leaked it would cause only mild contamination on the ground in the immediate vacinity of the spillage. It would be cleaned up by using detergents and water.[20]

Even though this ignores any possibility of radioactive gases or spent fuel constituents escaping (as assumed in the NRPB study), it does concede that at least one kind of radioactive leak is possible – the escape of contaminated water. Nonetheless, the leaflet concluded that "There is no evidence to suggest that there would be any long-term cancers resulting from a flask leaking". Given that it is now widely accepted that there is a finite risk of cancer from even the smallest dose of radiation, the implication that there would be no link between radiooactivity – even "mild contamination" – and cancer, represents public relations with a vengeance.

In any case, such evidence did exist. In 1978 – five years before the CEGB leaflet was published – the Political Ecology Research Group (PERG) had carried out its own analysis of a flask leak which reached very different conclusions. PERG made use of another UKAEA computer programme, called TIRION, to calculate the effects of a radioactive leak from a magnox flask in East London. However, unlike the later NRPB study, PERG's assessment assumed a release of 10 per cent of one particular substance – the radioactive fission product, caesium-137. The result was a hypothetical death toll of 597 fatal cancers. If all of the caesium-137 present in the flask were released, the death toll could rise to almost 6,000 – and that, it must be noted, was for just one radioactive substance; PERG's calculation took no account of the effects of releasing large quantities of any of the other constituents of spent fuel.[21] Greenpeace, for example, considered the possibility of a fire rupturing a flask of imported irradiated fuel near Barrow-in-Furness and concluded: "Tens of thousands of people would die of cancer over a 30-50 year period".[22]

The TIRION programme has since been criticised as unreliable, which is why the NRPB utilised MARC. However, MARC has also been used independently and with the input of different assumptions, it can produce dramatically different results. In 1987, the London campaign group ALARM (Alert Londoners Against Radioactive Material) commissioned a study to assess the effects of a magnox flask leak on the North London line in Hampstead. This study was, in effect, an update of the scenario postulated in the 1978 PERG report but using a more modern computer programme. Unlike its predecessor, ALARM's report considered the effects of *two* radionuclides, taking into account the release of caesium-134 as well as caesium-137. The report considered a range of possible effects: small, medium and high proportions of the caesium inventory being released, with a corresponding impact on public health. The least pessimistic forecast assumed a one per cent release of the caesium inventory with no resulting deaths. At the other end of the scale, however, was a 100 per cent release causing a death toll of over 100,000.[23]

Statistics like this lead to another area of disagreement – the likelihood of such a disaster occurring. Faced with concern about hypothetical accidents, the nuclear industry retreats into risk analysis. The likelihood of a serious transport accident is compared with the likelihood of non-nuclear accidents, just as the dangers of radiation are compared with other routine hazards in life. In this way, the transport of radioactive materials can be presented as an "acceptable risk".

In reality, however, it is questionable whether such statistics mean anything. Many people have the uncomfortable feeling that accidents are becoming more frequent. When major disasters actually happen, they are invariably regarded as "against-the-odds" events, so remote that no one thought they could ever occur. Recall the runway air collison of two jumbo jets near Tenerife or the chemical plant explosions at Flixborough and Bhopal: if these individual catastrophies were unpredictable, who could have anticipated the combination of transport accidents during 1987 and 1988 – the Zeebrugge ferry, the King's Cross fire, and the triple train crash near Clapham junction? These three tragedies were closely followed by two of Britain's worst ever air crashes: at Lockerbie and on the M1 near Castle Donington. Speculation about the cause of the Castle Donington crash threw up an interesting statistic. The likelihood of two engines coincidentally failing during the the same flight was reported as one in 10 million – exactly the same odds given by the CEGB for the probability of a serious accident to a spent fuel flask.[24]

As far as the nuclear industry is concerned, statistical predictions regarding the future are undermined by the industry's relative youth. Furthermore, history suggests that such statistics are invalidated by empirical experience. The 1979 accident at Three Mile Island provided a classic example. Prior to the event, the US Rasmussen report on reactor safety had predicted that the odds of such an accident occurring – that is, the odds of that particular sequence of faulty events happening in the way they subsequently did – were between one in 10 million and one in 100 million years of reactor operation – odds which seem to have been contradicted by the subsequent event. As the Scientific Advisor to Somerset County Council observed, recalling the Windscale Fire, Three Mile Island and Chernobyl before the start of the Hinkley C Inquiry: "Promises of an incredible 1:10⁶ or 1:10⁷ accident probability look unconvincing now that it can be said that a Beyond Design Basis accident has happened every ten years".[25]

Reports such as ALARM's, which predict such a wide range of potential effects, might be dismissed as meaningless. Yet common sense suggests that any comprehensive assessment of the possible consequences of a transport accident must recognise that the range of conceivable possibilities runs from the trivial to the catastrophic: at the bottom end of the scale lies the harmless bump, the minor traffic incident, or the derailment at walking speed in a railway siding. At the top is the "worst conceivable accident" where everything possible goes wrong – precisely the kind of scenario that the industry refuses to recognise.

15   The human factor

Various incidents have shown that the nuclear industry is vulnerable to human mistakes or misconduct. However sophisticated the equipment, however advanced the control systems, the industry ultimately depends upon people – even if the machines are perfect, their operators are rarely so blessed. Human beings are responsible for the design, construction and operation of everything from buildings and machinery, to computer programmes, nuts and bolts. Above all, people make decisions and people are fallible: all the major nuclear accidents to date – from the Windscale fire to Chernobyl – were directly or indirectly caused by human misjudgement. What has been described as the "worst radiological accident in the world"[1] – two years *after* Chernobyl – occurred at Goiania in Brazil, when a curious scrap metal dealer broke open a capsule of caesium-137 chloride from a defunct radiotherapy clinic. The release of 51.8 TBq of radioactivity resulted in several deaths and extensive contamination as people unknowingly spread radioactive material around. By accident or design, human actions can jeopardise the industry's operations and endanger the health and safety of the public.

The NRPB, for example, have reported that over half of the transport events causing increased radiation exposures between 1964 and 1988 were the result of wrongly prepared packages.[2] Amersham International have acknowledged the same problem: ". . . we have had a number of incidents or near-misses when customers have returned old radiographic radiation sources without making sure the container was properly assembled".[3]

The safety of transport containers could also be prejudiced by human errors made during construction, and not necessarily by people within the nuclear industry. In 1979, for example, the Boilermakers' Society alleged that some of the multi-element bottles used inside Excellox flasks had been incorrectly welded. According to former employees of Mersey Welding Ltd, a company which manufactured a batch of the bottles for BNFL, potentially dangerous short cuts were made in their construction, including welds that did not meet BNFL's specifications. Although BNFL admitted that some bottles had been returned to the company for "rectification", they denied that flask safety had been compromised.[4] A few years later, several magnox flasks were withdrawn after manufacturing faults had been discovered during maintenance work at Sellafield.

In other situations, human mistakes can result in radioactive materials getting lost. Examples in recent years include:[5]

- 1985:   An iron-55 source (200 kBq) goes missing from a scientific instrument en route to Finland from Clandon Scientific Ltd of Aldershot.
- 1986:   An iodine-125 source (2.25 GBq) reported lost by Lufthansa Airlines at Heathrow Airport.
- 1987:   An americium-241 source (11 GBq) lost "possibly from a bonded warehouse near Heathrow Airport".
- 1987:   A carbon-14 source (12 MBq) reported by Lufthansa Airlines to have gone astray at Heathrow Airport.

A more serious incident was recounted by Walt Patterson in his book *Nuclear Power*. Thirty three kgs of 90 per cent enriched uranium were mistakenly offloaded at London Airport during a journey from New York to Frankfurt. The fissile material – enough for a nuclear weapon – was left unattended until the company responsible for transporting it made enquiries to the airline.[6] While these incidents may have been inconsequential, the "human factor" includes a range of malevolent actions which could seriously endanger the transport and nuclear industries.

At one end of the scale, problems can be caused by theft and vandalism. Theft has been a minor problem in the nuclear industry as a whole. For example, unused uranium fuel elements have been reported stolen in the past from Bradwell nuclear power station and BNFL Springfields.[7] Cobalt-60 sources have been removed from several locations in Britain including the naval nuclear dockyards. Thefts have also occurred during transit when vehicles carrying radioisotopes have been stolen; possibly without the thieves realising what they were carrying. In January 1991, for example, thieves in Bootle stole a security van whose contents included a consignment of iodine-125.[8]

In 1989, vandalism caused the derailment of a nuclear flask train near Temple Mills in East London. The train went through a signal which would have shown red had its light bulb not been broken by vandals. The locomotive overturned and slightly damaged the flatrol; the flask itself was unscathed. The fact that health physicists from Bradwell nuclear power station arrived to check for contamination suggests that the incident was not taken lightly – minor acts of vandalism can precipitate serious consequences. At the other end of the scale, lie sabotage and the spectre of nuclear terrorism – unpredictable human factors which could expose the public to far greater potential dangers.

The human factor – terrorism

The nuclear industry operates in a world of political struggle, violence and war. Using raw materials which are radioactive and in many instances potentially deadly, it offers unique opportunities for disruption by people for political or personal reasons. Furthermore, the industry itself is often seen as an enemy – a juggernaut which invades local communities, contaminates the global environment, and whose products have on two occasions brought death to thousands of people. The threat of retribution, theft, or damage against anything nuclear – including radioactive materials in transit – are hazards which the industry cannot ignore.

Arguably two of the worst examples of what can be called "nuclear terrorism' were carried out by governments who in other circumstances are among the first to condemn such actions. In 1981, the Israeli air force bombed Iraq's Tamuz nuclear reactor at Daura just before it was due to start up, while during the 1991 Gulf War – when the west accused Saddam Hussein of "eco-terrorism" for firing Kuwaiti oil wells – the United States bombed Iraq's remaining two research reactors – both of which were fuelled with highly enriched uranium. But governments are not the only perpetrators of violence, as terrorists around the world have shown. Moreover, the inclination of some human beings, sane or otherwise, to wreak random murder and mayhem on their surroundings is a reminder that political struggle is not the only reason for violence. Potential threats to radioactive materials in transit might come not just from terrorists but from apolitical criminals or outright maniacs.

The large variety of radioactive materials moving around offers many opportunities for malign intervention, although some would be more tempting than others. Summarising the more dangerous potential threats is somewhat arbitrary, but four possibilities might be identified:

1 Theft of nuclear warheads.
2 Theft of fissile material for use in a nuclear weapon.
3 Theft of radioactive materials for subsequent dispersal, or threatened dispersal, to contaminate the environment or endanger life.
4 Attacks on containers transporting radioactive materials or warheads in transit to inconvenience or incapacitate part of the nuclear industry, or to release radioactivity into the environment.

The spectre of nuclear terrorism emerged as a public issue during the 1970s. In 1975, for example, the director of the US Arms Control Agency, Dr Fred Ikle, warned that the US had become "basically defenceless" against terrorist or criminal attack. Ikle argued that the international spread of sensitive nuclear technology had created opportunities for governments and other organisations to build nuclear devices either with the help of proliferating nuclear facilities or with "diverted" nuclear material, including fissionable materials stolen in transit.

In 1976, the hijack scenario was fictionalised in Britain by Chapman Pincher in his novel *The Eye of the Tornado* – a book which had a realistic eye for detail. Pincher described how nuclear warheads were transported every few weeks to Aldermaston from the Polaris submarine base at Faslane in Scoltand. Part of the journey, from Gare Loch to Portsmouth, was made by sea using a Royal Navy Auxiliary ship, the *Mallard*. In Pincher's story, the Mallard's course passed close to the Irish Coast, where, on one particular sailing, the *Mallard*, with ten Polaris warheads on board, stopped for a vessel in distress. Unfortunately, those rescued by the *Mallard* from the sinking ship turned out to be IRA hijackers who were part of an international communist conspiracy. By opening up the warheads and bypassing the arming mechanism with a few bits of wire, Pincher's left-wing bogeys threatened to detonate the weapon and contaminate southern Britain with radioactive fallout blown from the Irish Sea, unless their demands were met.

Allegations of a suspected real life hijack were made in the same year. Two Hungarian envoys were arrested by police after photographing the Royal Ordnance Factory at Burghfield. The incident provoked much speculation about their intentions. Had their real interest, perhaps, been the movement of warheads or their nuclear components between Burghfield and Aldermaston? The *Sunday Express* – which coincidentally had been serialising Pincher's book – reported that British security chiefs believed that the Hungarians were planning a spectacular nuclear hijack, backed by Soviet bloc intelligence: "Those leading the security inquiry . . . think that it was part of a strategy to establish the pattern of convoys carrying nuclear cargo and to learn the strength of the guard on them".[9]

However disastrous the theft of a warhead might be, it is not necessarily the most likely terrorist scenario. Warheads are transported amid stringent military security and, in contrast to their police escorts, the devices themselves are unarmed. Notwithstanding the ingenuity of KGB-controlled Irish Marxists, other radioactive materials would be relatively easier to steal and in many ways more attractive targets. The expansion of civil nuclear power has increased the movement of potentially lethal materials like plutonium and spent fuel. The material movements generated by fuel reprocessing have been a particular cause for concern. During the 1977 Windscale Inquiry into the proposed new THORP reprocessing plant, even BNFL's managing director, Mr Conningsby Allday, accepted that the theft of plutonium by terrorists who could manufacture a crude nuclear bomb, was a possibility.[10] As the amount of nuclear material moving around increases, so must the risks of interference.

The vulnerability of radioactive materials in transit was highlighted dramatically by the "Plumbat affair" which came to light in 1977. On 17 November 1968, a cargo ship, the *Scheersberg A*, sailed from the Belgian port of Antwerp with 560 drums of yellowcake. The 200 tonnes of uranium had been bought by a German chemical firm from the Societe Generale des Minerais, a Brussels-based company connected to Union Miniere which once owned the Shiknolobwe uranium mine in Zaire. The *Scheersberg A* was ostensibly bound for Genoa in Italy, but it never arrived. Instead, on 2 December 1968, the ship turned up in Iskenderum in Turkey – minus its nuclear cargo. Strictly speaking, the destination of the uranium remains unproven. But the fact that the German company – Asmara Chemie of Wiesbaden – was owned by a Jewish family and frequently acted on behalf of Israel tied in with the revelations of an Israeli secret agent arrested five years later in Norway: somewhere in the Mediterranean between Turkey and Cyprus, the *Scheersberg A*'s cargo had been offloaded – ending up in Israel to fuel its plutonium-producing reactor at Dimona.

The disappearance of the *Scheersberg*'s uranium was a "diversion" rather than a hijacking. First revealed by the US magazine *Rolling Stone*, it drew attention to Israel's clandestine efforts to manufacture nuclear weapons. The magazine also alleged that in the late sixties, agents of Mossad, the Israeli intelligence service, set up a special commando unit to raid the Western nuclear powers. On several occasions, Israel successfully obtained supplies of enriched and unenriched uranium from European and US sources. In all cases, the incidents were hushed up by the victimised governments to avoid political embarrassment. For example, in the same year as the Plumbat affair, Israeli agents allegedly tear-gassed the driver of a lorry transporting uranium and drove the vehicle and its cargo to the Negev desert, location of the Dimona reactor. A short time later, the operation was reportedly repeated somewhere in England and a captured consignment of yellowcake pirated back to Israel.[11] Predictably, the magazine's allegation's were firmly denied by the Israeli Foreign Ministry as "dramatic and fantastic stories . . . which have no foundation in reality" but, as the magazine pointed out, embarrassed western governments were equally unlikely to acknowledge such events.

Stealing a consignment of plutonium or highly-enriched uranium – materials from which warheads can be constructed – might be a tempting alternative to filching an actual warhead. The amount required for a nuclear weapon is defined by the IAEA as a "significant quantity" (SQ): either 8 kilograms of plutonium or 25 kgs of highly enriched uranium. Whether a determined individual could build a workable nuclear warhead is debatable. Nevertheless, several simulated attempts have been made to prove the point and it might be assumed that ambitious governments would find the task even easier.

Plutonium, more widely used in the civil nuclear industry, might be regarded as being at greater risk of theft. With the expansion of reprocessing, the increasing use of mixed-oxide reactor fuels, and the development of fast breeders, the amount of plutonium in transit has been steadily increasing. Despite the secrecy surrounding shipment times, any intelligent criminal could figure out where plutonium might be found from the locations of the relevant fuel cycle facilities – no amount of secrecy can hide the fact that the transport routes run between them.

Consignments of plutonium metal might well be the first choice for a thief. Other forms of plutonium, such as plutonium nitrate, would require hazardous but chemically straightforward processing before being suitable for a warhead: plutonium nitrate can be converted to plutonium dioxide by simple precipitation; plutonium dioxide can be used directly in an unsophisticated warhead. Theoretically, such operations need not be beyond the capability of a resourceful criminal with a suitably equipped laboratory.

Theoretically, plutonium might be obtained from stolen spent fuel. However, even if it were possible to steal a 50-tonne flask of the material, the successful thieves would have the near insuperable task of extracting the plutonium from its contents. In fact, the chemistry of reprocessing is not particularly secret. Patents available from the Patent Office in London give any would-be rival to BNFL a choice of chemical methods. Nevertheless, the possibility of do-it-yourself reprocessing is usually regarded as too problematic to be feasible, if not pure science fiction. The intense radioactivity would be lethal to anyone not massively protected by concrete or lead. In any case, the amount of plutonium inside a single loaded spent fuel flask is usually less than the quantity needed for an unsophisticated warhead – unlike consignments of plutonium nitrate or fast reactor fuel which would usually have more than enough.

In practice, however, it need not be necessary actually to build an explosive device. Proof that fissile material had been acquired or evidence that the blackmailer could build a bomb – perhaps by releasing authentic designs or convincing technical information – might be enough to validate the threat.

Alternatively, fissile material, or any other dangerous radioactive substance, might be stolen for threatened dispersal – an obvious opportunity for a ransom demand. A liquid like plutonium nitrate or solid plutonium dioxide in a powdered form would be ideal. Such threats have already occurred. In 1985, several New York City officials received anonymous letters threatening to contaminate the City's water supply with plutonium. Moreover, subsequent measurements of water samples by the US Department of Energy found higher than normal readings of plutonium. ("Normal" readings arise because some three tonnes of plutonium has been deposited on the planet by nuclear weapon tests.) The conclusion, however, was that the threat was a hoax and the plutonium had arrived by some other means. (One suggestion was that an animal had spread contamination from a disused plutonium handling facility.)

A earlier threat was for real. In 1979, two steel drums containing 145 pounds of enriched uranium were stolen by an employee at a General Electric plant in Burlington, New Carolina. The employee threatened to disperse the uranium over two American cities unless he received money from the company. The thief was caught before the threat could be carried out.[12]

Sabotage would undoubtedly be easier than theft. There has been a recurrent fear that spent fuel flasks in particular – which transport massive amounts of radioactivity – might be attacked simply to release their contents. The possibility is recognised by the industry itself. In 1985, for example, the UKAEA's Safety and Reliability Directorate noted: "In order to ensure safety . . . flasks are designed to withstand damage occurring to them should they be involved in an accident. Thus existing safety measures should prevent any problems arising with the transport of flasks. *There are, however, some extreme conditions which are not covered and these include sabotage.*" [Author's emphasis.][13]

One scenario was demonstrated in East London by the Freedom of Information Campaign. The well-publicised event has been described by one of those involved: "On Halloween night, 1979, I walked onto Stratford Station with a Rocket Launcher on my shoulder, waited on platform nine in full view of British Railway's staff until a train carrying a 51 tonne flask of spent nuclear fuel pulled in. I then calmly aimed the launcher at the flask and pulled the trigger . . ."[14] The event was immortalised by British Rail's memorable response: they had every right to be on the platform as long as they had a platform ticket; nor was it the job of British Rail staff to apprehend people carrying rocket launchers – which was of course a harmless replica.

In contrast to British Rail's apparent indifference, the government took the issue more seriously. A special study was commissioned into the vulnerability of flasks to terrorist attack. According to Energy Minister John Moore, the study, completed in 1982, confirmed the safety of flasks: ". . . even under a most adverse combination of circumstances these flasks would not give rise to any significant hazard to the local population. I am satisfied that the existing arrangements . . . are adequate to protect public safety against any consequence of such an attack and that there are no grounds for altering them."[15] Needless to say, the government decided for security reasons not to publish the study – a better indication perhaps of the report's sensitive conclusions.

Bomb threats and sabotage are not unknown to the nuclear industry. Reactors worldwide – working or under construction – have been targets for sabotage, actual and threatened. Radioactive materials in transit are also potential targets. During the 1982 Dublin trial of Gerard Tuite, an alleged IRA member charged with possessing explosives, the prosecution played a tape-recorded "hit list" which identified the Cumbrian coast railway lines to Sellafield as potential targets.[16]

The British nuclear industry has a special significance in Ireland, whose government has persistently opposed radioactive discharges into the Irish Sea from Sellafield. To Irish nationalists, BNFL's radioactivity is yet another unwelcome British intrusion. As trains of spent nuclear fuel provide raw material for Sellafield's pollution, bombing the isolated stretch of Cumbrian coast facing the Republic would have had a tactical as well as a symbolic significance to militant Irish nationalists.

IRA threats against the nuclear industry could hardly have been a surprise. Other industries had been targets of IRA bombs and the paramilitaries had a well-known interest in railways. Stations in Northern Ireland and Britain have been repeated targets for attacks as have the trains themselves. In October 1978, for example, three bombs were planted on board a Dublin to Belfast passenger train. All three exploded – the first while the train was approaching a station on the outskirts of Belfast. Six months later, a freight train on the line from Dublin to Belfast was hijacked, loaded with explosives and blown up. In 1980, in an attempt to sever the link between the Republic and Northern Ireland, the IRA claimed that it had used half a ton of explosives to blow up the Kilnasaggart railway bridge, a target of repeated attacks.[17] More recently (25 February 1991), the IRA were held responsible for detonating a high explosive device on the Midland line south of St. Albans. A metre-length piece of rail, flung into the air by the force of the blast, sliced the roof off a car as it landed.

Any review of terrorism in recent years will confirm that transport systems are targets. All modes of transport have been attacked. Ireland is not the only country where railways have been bombed. Two days before Christmas 1984, 15 people were killed and over 150 injured in Italy when a bomb exploded on a Naples to Milan passenger train as it sped through a tunnel in the Apennine mountains near Bolgna. A decade before, 12 people had died when another train was bombed in the same tunnel.

Bombs on aircraft are well-known, although it might be assumed that security on the special flights now used for materials like plutonium would preclude a repetition of a Lockerbie-type disaster. For the industry, the greatest danger in these situations is the "insider threat" where employees turn malevolent, perhaps in cahoots with outsiders. Yet aircraft are also vulnerable to attacks from the ground. In 1976, for example, factions of the Popular Front for the Liberation of Palestine were arrested near Nairobi airport while preparing to fire SA-7 surface-to-air missiles at an incoming El Al jet. Two years later. the Soviet-made SA-7 was suspected of hitting a passenger aircraft in Zimbabwe which crashed killing 38 people.[18] Transporting plutonium by air is no quarantee of absolute security.

As for sea transport, one of the earliest known losses of radioactive material in transit occurred during the second world war. German submarines sank two ships and a total of 400 tonnes of uranium ore concentrates on their way to the United States from the Belgian Congo (although the Germans would presumably not have known what the ship was carrying).[19] More recently, shipping routes used by flasks of irradiated fuel have been threatened with attack. In 1986, a suicide squad was reportedly training to attack targets in Latin America, including the Panama Canal – through which Japanese fuel is shipped to Europe.[20] In the same year but closer to home, police were alerted in Dover after a tip-off about a plot to blow up and sink a ferry in the English Channel.[21]

It must be assumed that the possibility of nuclear terrorism in Britain is recognised by the British government, although officially the subject is taboo or at least downplayed. For example, Francis Tombs, when chair of the Electricity Council, regarded the theft of nuclear materials for weapons as: ". . . a hazardous and uncertain procedure involving considerable risks for the terrorist with unpredictable and long drawn out results. One wonders why an intelligent terrorist should undertake such a task when there are readily available much more certain and less dangerous ways of blackmailing a population".[22]

In contrast to Britain, terrorist threats are more openly acknowledged in the United States, and with good reason: by 1986, more than seventy threats of nuclear terrorism – including nuclear bomb threats – had been investigated by the FBI. All but one (the theft of uranium from General Electric, cited above) were hoaxes, including an occasion in 1974 when an extortionist threatened to explode a nuclear bomb in Boston unless paid $200,000. This particular incident revealed deficiencies in the authorities' reactions and led President Ford to create a special response organisation, NEST – the Nuclear Emergency Search Team. Since then, NEST has responded to nuclear bomb threats in various cities including Los Angeles, San Francisco and New York.[23] The fact that NEST has set up a branch in Europe – based at the US air base at Ramstein near Frankfurt – indicates that the threat of nuclear terrorism is not seen as unique to the States.

The spectre of terrorism presents a dilemma to those concerned about the transport of radioactive materials: is a discussion of the threat not likely to exacerbate the danger? The risk of malicious interference is one reason why the nuclear industry keeps many of its operations so secret. Perhaps if the industry were totally isolated from the rest of society, this would not be controversial. But nuclear power affects everyone in one way or another – some of us use its electricity; all of us consume its pollution. Many would also be affected by acts of nuclear terrorism.

16   Secrecy

On 15 February 1989 a train of empty Excellox flasks arrived at BNFL's berth in Barrow, where the *Pacific Sandpiper* waited to sail to Japan. This was not a typical journey. In addition to the empty flasks returning for more fuel, the train carried two full flasks of irradiated Japanese research reactor fuel bound for reprocessing in the United States. The fuel had arrived in Britain six months earlier, and been stored temporarily at Sellafield. Such shipments are made on average about once a year. The revelation that BNFL were transporting Japanese research reactor fuel to the United States proved something of an embarrassment for the company, as such shipments appeared to breach a US embargo on research reactor fuel imports, imposed while transport safety was being assessed.

But there was also something else on the train. Two of the wagons each carried a pair of cylinders resembling oversized beer kegs. About two metres tall, the kegs were framed with a mesh of metal bars not unlike those found behind a fridge – pecisely the sort of container that might be used to transport tritium. Although they were seen by people in Barrow, BNFL denied their existence. The company told the local authority that "the consignment which departed from Barrow on 21 February consisted entirely of irradiated fuel elements". At that time, tritium production in the States had stopped and it would have made obvious sense for the US government to import stop-gap supplies from Britain, presumably under the Mutual Defence Agreements. In the absence of any other official explanation, the fact that BNFL denied all knowledge of the cylinders inevitably raised suspicions.

The incident demonstrates that, despite the industry's desire to appear open and informative, secrecy is still endemic. The reasons are obvious: military matters are protected for "national security" while contracts and business affairs are "commercially confidential". Thus one of the most important regulations affecting the transport of radioactive material is the Official Secrets Act. Most people in the industry have to sign the Act, which makes the unauthorised disclosure of certain information an offence. Although the Act applies mainly to the industry's military activities, civil operations, including transport, are also affected as the Energy Secretary has explained:

> Some atomic energy information within the civil field is classified because it relates to the technologies for or the scale and cost of production of those nuclear materials used in nuclear weapons; and to the storage and transport of such materials. Additionally, classification is necessary to honour the United Kingdom's international and treaty commitments to non-proliferation policies and to prevent disclosure of information of potential value to those who might seek to acquire fissile materials illegally or to damage installations or materials in transit.[1]

Presumably this is why the Town and Country Planning Association (TCPA) fell foul of the Act when it presented evidence to the Sizewell B Inquiry in 1984. Part of the Association's case concerned the vulnerability of spent fuel flasks to paramilitary attack with "common explosive devices". The TCPA's witness, Marvin Resnikoff from the US Radioactive Waste Campaign, described how a flask could be pierced with a conical-shaped charge and, with the help of an incendiary pellet, "fragment fuel rods and pellets, vaporise semi-volatile radionuclides such as caesium and release radioactivity from the flask due to overpressure".[2] Despite the fact that such explosive devices are used by – among others – the oceanographic, construction and steel industries, the Inquiry inspector, Frank Layfield, ruled that Resnikoff's evidence contravened the Official Secrets Act and could not be read into the Inquiry transcript.

The Official Secrets Act is augmented by voluntary restrictions on the media known as D Notices. These are issued by the Defence Press and Broadcasting Committee, chaired by the Permanent Under-Secretary of State for Defence. The committee comprises officials from government departments concerned with national security and representatives of press and broadcasting organisations. D Notices apply to information covering the security services, defence plans and civil defence, which the government prefers not to be published because it would be "harmful to the nation". Although D Notices have no legal authority, they serve to remind editors that publication of certain information could contravene the Official Secrets Act. Of the eight D Notices existing at the time of writing, No. 3 is entitled "Nuclear weapons and equipment". The information that the committee requests should not be published "without first seeking advice" includes:

> stockpile quantities, storage arrangements and deployment plans for nuclear weapons and the materials used in them, especially plutonium, uranium and tritium" and "detailed information on the arrangements for transport and movement of nuclear weapons, components and fissile material.

Yet official secrecy – however repugnant to libertarians – is at least understandable; it is the habit of militaristic governments to guard their military secrets. Less understandable is the secrecy that surrounds the movement of many less sensitive radioactive materials. The refusal to supply information often appears arbitrary, inconsistent and confusing – especially when information is released in one context but withheld elsewhere. The case of Rolls-Royce refusing to tell Derbyshire County Council information that the company themselves had previously submitted to – and which had been published by – a House of Commons Select Committee is a classic example (see Chapter 5). When information is already in the public domain, it is clear that secrecy is less a legal requisite than a ploy to avoid giving information to people who might question, embarrass or otherwise potentially cause trouble for the industry's operations: local authorities, environmental groups, trade unions, councillors, MPs, journalists, and not least, the general public.

Sometimes the failure to supply information looks like ignorance. Time and again members of parliament have asked the Department of Transport for details of radioactive materials passing through their constituencies, only to be told that "the Department does not collect information of that kind" or words to that effect. Such replies from a government department whose job is to regulate the transport of radioactive materials suggest an abdication of its responsibilities. If the Department of Transport appears not to know what's moving around Britain, it can hardly be surprised if members of parliament have doubts about safety.

In some instances it is clear that the department could supply the requested information but chooses not to. In October 1989, for example, Labour MP Martin Redmond asked the Transport Secretary how many tonnes of radioactive material had been imported annually during the previous ten years through various British ports such as Barrow and Dover. The reply was typical: "The Department does not collect this information . . ."[3] Yet less than six months earlier in response to a similar parliamentary question about specific ports on Humberside, the same department had supplied detailed figures showing precisely how many tonnes of different types of radioactive material had been imported and exported through Immingham and Hull for each of the previous ten years.[4]

In the following year, Liberal MP Simon Hughes asked the Transport Secretary simply to list the ports through which radioactive materials had been imported or exported – an innocent request for general information that would surely give away no state secrets or commercial confidences. But the answer? "The Department does not collect data of the kind requested."[5] If that is not mere evasion, then it must be interpreted as a glaring deficency in the department's responsibilities.

Other examples show that sometimes the department is indeed ignorant of its subject. In 1990, for example, the Transport Secretary was asked how much low level and other types of "nuclear" waste had been transported by train since 1979. Virtually all movements of low level waste have been between Sellafield and BNFL's railway sidings at Drigg. Ignoring this, the Transport Secretary stated ". . . the only movements of waste were of intermediate level . . ." (referring to the years when waste was dumped at sea).[6] On another occasion, Oxford MP Andrew Smith asked whether any uranium passed through his constituency by rail; the Transport Secretary replied that normally it didn't,[7] despite the fact that yellowcake imports from Southampton regularly pass through Oxford on Freightliner trains to the north of England (see Chapter 3). Such disinformation can only be interpreted as ignorance, indifference or worse.

The general lack of openness in Britain is highlighted by comparisons with other countries. The availability of information about shipments of radioactive material, transport routes and flask testing all provide examples of different approaches.

In Australia, for example, the anticipated export of spent fuel from the Hifar research reactor near Sydney to the United States for reprocessing prompted the Australian Nuclear Science and Technology Organisation (ANSTO) to mount a public relations exercise to inform the community about the proposed operation. ANSTO's efforts included writing to all local councils along the likely transport routes to ensure they were informed and had an opportunity to ask questions. A press conference was held to inform the media and meetings were organised with transport and waterfront unions to inform and discuss the operation. ANSTO even offered Greenpeace Australia access to the Safety Analysis Report on the transport flask to be used.

In the United States, the Murkowski amendment on air shipments of plutonium stipulated that the NRC "shall provide for public notice of the proposed test procedures, provide a reasonable opportunity for public comment on such procedures, and consider such comments, if any". This contrasts sharply with British practice. While the CEGB's magnox flask test programme in the early 1980s was conducted with massive publicity, tests on other types of package have not had the same high profile – indeed, quite the reverse. BNFL have refused to allow journalists to witness their tests on plutonium containers (to see if they met US NRC standards) while other test results are kept secret. For example, test results for flasks used for irradiated research reactor fuel are, according to the Transport Secretary, "the property of the UKAEA and supplied to the Department on a commercial-in-confidence basis".[8]

The United States provides more examples of relative openness. As a nuclear weapon state with a large civil nuclear industry as well, it has just as much reason to cite "national security" or "commercial confidentiality" for being secretive. Yet the availability of information on many aspects of the US nuclear industry serves to remind how much is withheld in Britain.

At the less sensitive end of the range, published US trade statistics show the quantities of uranium oxide and uranium hexafluoride exported to Britain, where statistics of the same materials arriving as imports are deemed unsuitable for publication.

Import and export licenses provide a further contrast. In the United States, the NRC publishes details of applications for licenses to export nuclear materials, including plutonium, tritium (for industrial uses) and highly enriched uranium. These are reported in trade journals and are available from the NRC itself. The Commission will supply, for example, a list of licenses granted for the export of highly enriched uranium to Britain for civil use over the last 20 years. In Britain, the government will not even reveal the total number of export licenses granted.[9]

At the other end of the scale, there is significantly more openness in the US about the movement of fissile materials. The annual reports of the Nuclear Regulatory Commission (NRC) even publish photographs of plutonium consignments in transit,[10] something which Britain's Department of Energy or BNFL seem unlikely ever to do.

A US Department of Energy Safe Secure Transport vehicle used for moving nuclear warheads and special nuclear materials

Elsewhere details can be found of the vehicles used to transport fissile materials. The US government's Sandia Laboratories have published details of the nationwide safe-secure transportation system used in the United States for the road transport of highly enriched uranium and plutonium.[11] The system is centred around a fleet of Safe-Secure Trailers (SSTs) – also known as Safe Secure Transport vehicles – introduced by the US government's Energy Research and Development Administration (ERDA) in 1972. The SSTs are similar in appearance to a standard semi-trailer but with walls, ceiling and floor specially constructed of impenetrable materials that will resist attack for an extended period of time.[12] They are accompanied by special escort vehicles and utilise a dedicated nationwide communications network called SECOM. This keeps vehicle convoys in continuous contact with the ERDA control centre in Albuquerque, New Mexico. But how different the attitude on this side of the Atlantic: "It would not be in the public interest . . ." again and again ad nauseam.

The nuclear state

Measures to prevent the disclosure of information are not confined to the statute book. The nuclear industry and the government take active steps to ensure that their operations are kept private.

Take John, for example – not his real name – a railway buff from Derby. Driving along Raynesway one Saturday he stopped and photographed a nuclear flask on the railway line that crosses under the road, coincidentally near to Rolls-Royce. Within minutes the police arrived and told him to put his camera away. That in itself is not unusual, but he didn't expect what followed. Just before midnight the next day, two men identifying themselves as "government security agents" turned up at his house and questioned him for more than an hour. John's interest in photographing railways convinced them he was not a spy. Nevertheless, when his film returned from the processors, the four shots he had taken of the flask had mysteriously failed to come out. On other occasions photographers snapping nuclear convoys have simply had film ripped out of their cameras.

Such actions are to be expected near military establishments. The chain link fence around Aldermaston, for example, displays large signs warning that photography and sketching is forbidden. MoD police have specific instructions regarding people seen photographing the site from outside: they are to suggest to them ". . . that it may be in their interest that the film from the camera be removed . . . in order that the film may be developed to prove their intent was not prejudicial to the interests of the State". When similar restrictions are applied to materials in transit – whose military purpose is not always apparent – then official secrecy is potentially all around us.

The potential threat of terrorism or theft provides one justification and is also a reason why the UKAEA has its own armed police force, one of whose duties concerns the transport of radioactive material:

Constables carry firearms when they are on duties related to the guarding
of special nuclear material on sites or in transit . . . armed constables are
deployed on the AEA sites at Dounreay, Harwell and Winfrith and the British
Nuclear Fuels (BNFL) site at Windscale. However, it would not be in the
public interest to give details of the deployment of AEA constables on these
duties either at sites or in transit.[13]

If secrecy were only about the prevention of terrorism it might be regarded
as acceptable. But unfortunately it conflicts with other interests, and
especially with health and safety. Is public confidence in the transport
of plutonium nitrate helped by knowing that, for reasons of security,
coastguards are not informed of the shipments? The locations of spent fuel
flask "layovers" provide another example. Layovers are timetabled breaks in
the journey to Sellafield when flasks are parked in railway marshalling yards
or sidings for periods of up to half a day. Such locations might be potential
targets for malicious attack; flasks are stationary and access is easy. Yet
the locations concerned are also responsible for the highest radiation doses
received by the public from the transport of irradiated fuel. While it may be
in the interests of a debate about public health to reveal the locations, such
information could be useful to terrorists.

Secrecy is not just objectionable as a matter of principle. In the past, it
has been used to cover up problems which become publicly known only
years afterwards when the secrecy is broken. Various revelations about the
early days of Sellafield illustrate this unfortunate habit. Not until 1983, for
example, was it officially acknowledged for the first time that the Windscale
fire had released quantities of radioactive polonium, a disclosure which
subsequently changed official estimates of the number of resulting fatalities.

Examples of accidents and problems affecting the transport of radioactive
materials are also beginning to surface after years of being kept under wraps.
In 1948, for example, a lorry transporting uranium between Rock Savage
(the ICI chemical works in Runcorn) and Springfields caught fire. The fire
charred the lorry's floor decking and was referred to in a 1960 report on fire
tests for plutonium containers. The fire test report was kept secret and only
when it was made available twenty five years later in the Public Records
Library did knowledge of the earlier accident emerge.[14]

Another once secret report by Harwell in 1956 investigated repeated
cases of contamination on lorries transporting radioactive materials between
Harwell, Windscale and Amersham. The materials transported included
spent fuel from Harwell's BEPO reactor. Contamination was found on
the undersides of the lorries, on their wheels, and even inside the drivers'
cabs. Beta/gamma radiation levels of up to 2000 cps (counts per second)
were found on one part of a lorry and on a worker's PVC clothing.[15]

These incidents may or may not have been serious. But the fact that they have emerged after being deliberately withheld makes one wonder what else has been hidden. The secrecy of the past casts doubt on current practice. To put it another way, will we have to wait thirty years from now to find out what has really been happening in the nuclear industry of today? When the industry and the government continue to withhold information, the suspicion remains that something is being covered up.

17   Emergency planning

Contingency plans for dealing with accidents to radioactive materials in transit are prepared by the industry itself. This is a legal obligation under regulation 27 of the Ionising Radiations Regulations 1985. Some plans are intended for particular materials or operators; others have a wider application. There are also differences in the extent to which their details have been made public.

For incidents involving irradiated fuel flasks, Nuclear Electric and Scottish Nuclear maintain "The Irradiated Fuel Transport Flask Emergency Plan", a summary of which has been published by the companies.[1] The plan specifies arrangements for obtaining specialist assistance which, in the case of railway accidents, would be activated by the train crew through the appropriate railway control office or, for mishaps on a road, through a local nuclear establishment.

The plan prescribes the principal duties of the parties involved. British Rail would notify Nuclear Electric's Alert Centre in London (Scottish Nuclear in Scotland and additionally BNFL in southern Scotland) and whichever emergency services were required. The Alert Centre would call out a Flask Emergency Team from a nuclear power station who would examine the flask, carry out radiation monitoring in the surrounding areas and, if required, call for additional assistance.

Major accidents involving other radioactive materials would be dealt with by NIREP – the "UK Nuclear Industry Road Emergency Plan" drawn up by the major operators in the nuclear industry. The plan covers packages owned by Nuclear Electric, Scottish Nuclear, BNFL, the UKAEA, Amersham International, Rolls-Royce and Associates and the Ministry of Defence. Excluded are Excepted packages, nuclear weapon cargoes and irradiated fuel flasks travelling by road to or from a railhead.

NIREP is intended to provide a rapid response to any incident (other than a routine vehicle breakdown) involving radioactive materials transported by the participants. This includes deliberate attempts to "damage or interfere with a package". In the event of an incident, the vehicle driver or his or her "mate' is expected to notify the UKAEA constabulary's Force

Communications Centre (FCC) at Risley. The FCC was set up by the UKAEA police to monitor and assist communications for the transport of radioactive materials between Authority sites and BNFL. It is staffed by personnel from the UKAEA constabulary who would initiate appropriate action. The FCC would call out assistance from the nearest nuclear site of the one of the participating orgainisations. This comprises a health physicist, with radiation monitoring equipment, who would examine the package and check for contamination. The "responsible site" – usually the consignor of the radioactive material – would be expected to take over the task of dealing with the consequences of the accident. Meanwhile the driver must report the incident to the local emergency services and, if there is evidence of a leak or a fire, keep the public at a safe distance from the package.

NIREP is similar in many ways to the NAIR scheme – the National Arrangements for Incidents Involving Radioactivity – which has been available for many years to deal with minor accidents involving radioactive materials in the public domain. The NAIR scheme provides nationwide emergency assistance, available through the civil police, for incidents where radiological hazards arise away from establishments where radioactive materials are regularly handled, such as nuclear sites or hospitals. It is co-ordinated by the NRPB and is available for transport accidents not covered by other contingency plans.

Assistance under the NAIR scheme is provided in two stages. Stage 1 assistance is available from 61 NHS hospital medical physics departments and various nuclear sites and research establishments. Under this arrange-ment, a qualified health physicist turns out to check whether radioactivity has been released which could create a risk to the public. If a hazard exists, advice is given on action to eliminate it or contain it until further assistance can be provided. Stage 1 assistance is obtained by informing the local civil police of the incident and asking them to contact the nearest establishment providing this service. If necessary, more specialised help can be called in to deal with more serious radiological incidents. This is Stage 2 assistance, available from the major nuclear establishments.

Since its introduction in 1964, NAIR has responded on average to approximately 15 incidents a year, few of which have been radiologically hazardous. Many involved undamaged radioactive sources which had gone astray and subsequently turned up where they shouldn't. Others included cases where problems arose while sources were being used in public places, for example, for industrial radiography. Some turn out to be false alarms – including toys marked with realistic radiation symbols – while others are hoaxes. Although the NAIR scheme was originally established to deal with transport incidents these comprise only a small proportion of call-outs. Most, such as a box containing twenty-five smoke detectors falling off a lorry crossing the Severn Bridge, have posed little danger to the public.

It goes without saying that plans for dealing with accidents to military nuclear materials, including nuclear warheads and nuclear submarine fuel, are secret.

Criticisms of emergency plans

Dealing with accidents has been a major area of concern, not least for local emergency services who would expect to attend. Any loss of shielding or containment could expose rescue workers to invisible and deadly dangers. Home Office guidance to the fire service gives an idea of the hazards they could face if attending an accident involving nuclear warheads:

> Fire fighters should approach the accident site from an upwind direction wherever possible with radio silence imposed . . . if a fire is still burning at the accident site prompt and decisive action will be required to prevent releases of radioactivity or further releases of radioactivity occurring . . . if the weapon is jetting bright white (flaming under pressure) an explosion of conventional HE could be imminent . . . any cooling jets should be lashed if possible and personnel should take protective cover.

Hazards to fire fighters are made explicit in the Nirep plan, which notes that although the dispersal of radioactive material is unlikely, "The incidents most likely to lead to such dispersal are serious vehicle fires". Much attention has focused on the transport of irradiated fuel. Both the Fire Brigades Union (FBU) and the National Union of Public Employees (representing ambulance workers) have expressed concern that their members might be exposed to unacceptable doses of radiation if they were called to an accident where a flask had been breached. In the 1980s, both unions passed resolutions at their annual conferences criticising existing arrangements. The FBU even went so far as to totally oppose the transport of irradiated fuel from "every existing or planned nuclear power station or installation in the country".

It is not difficult to see weaknesses in the industry's contingency plans. Both the Flask Emergency Plan and NIREP rely upon the train crew or vehicle driver to initiate emergency action; if the driver or crew were dead or incapacitated, the first line of communication would be disrupted from the start. The NAIR scheme has also had problems. On at least two occasions the NRPB has expressed concern that some members of the constabulary – who are instrumental in initiating NAIR action – seem to be unaware of the plan's existence.[2]

Even if an accident response proceeded as planned, there would inevitably be a time-lag before specialist assistance arrived. Various incidents in London over the years involving suspected leaks from flasks (all of which were subsequently declared safe) have shown that it takes several hours for a Flask Emergency Team to arrive in the capital.

Delays can occur even near a nuclear site. In May 1991, for example, BNFL staged an exercise in Cumbria to test the emergency response to a simulated accident. The accident scenario assumed that a vehicle loaded with inflammable liquid had crashed into an Excellox flask on a level crossing and caught fire. The exercise was staged in a railway siding in Millom, using a van parked by an empty flask. Although there was no actual crash or fire, the results were not encouraging. Apart from highlighting the local fire brigade's poor radiation monitoring equipment, it took the BNFL team leader – in an AEA police Land Rover – one hour and twenty five minutes to get to Millom from Sellafield, a distance of just over twenty miles. A BNFL "crash wagon" took two hours to arrive at the scene. Challenged about the delay after the event, BNFL blamed British Rail for not informing them; British Rail's excuse was that they had also been dealing with two real life (but non-nuclear) incidents in the area on the same day – a bomb scare near Carlisle and a fire at Grange-over-Sands.

It is an inherent problem in devising any contingency plan that a transport accident can happen almost anwhere. Thus if a major radiation leak occurs, the inevitable delay before specialist help arrives could result in people being irradiated and radioactive material dispersing into the environment.

The role of local authorities

The uncertainties about the effects of a major transport accident and how it would be dealt with are nowhere felt more acutely than in local government. Housing, transport, schools and parks could all be affected or put at risk if radioactivity were released. Yet while local authorities have an obvious concern for their community interests and people, their precise responsibilities in the general event of a disaster have not always been entirely clear. According to the Home Office, local authorities are required to maintain their own services, help people in distress, and co-ordinate what is being done by the various organisations which are giving help.[3] As the Association of Metropolitan Authorities has pointed out: "there is no general statutory duty on local authorities to plan for major peacetime emergencies or disasters".[4] Such plans are discretionary and widely regarded as inadequate or non-existent. Those which do exist are often rudimentary by-products of civil defence preparations which councils are legally obliged to make.

For radiation releases in general, there are several responses – notably evacuation and sheltering – which might reduce or avoid a dose to the public and would involve the local authority. Sheltering is most effective when radioactive material is being dispersed. It reduces exposure to airborne radioactivity and the chance of inhaling radioactive particles. Evacuation is a way of avoiding a post-accident dose from any subsequent ground contamination. Moving people in the immediate vacinity of an exposed source of gamma radiation would also reduce exposure.

The need for countermeasures would depend on the severity of the accident: it has been estimated that a catastrophic leak from a nuclear flask at Earl's Court in London could *in extremis* require the evacuation of 80,000 people for periods ranging from months to 125 years.[5] But if such a situation arose, who would decide what action was needed and who would carry it out?

An NTL flask transporting spent fuel from Goesgen nuclear power station, Switzerland, passing through London on its way to Sellafield. An accident on this stretch of the West London line could, according to some critics, require widespread evacuation.

Local authorities are responsible for the preparation of evacuation plans whose implementation is the job of the police. Yet when it comes to predicting the consequences of accidents involving radioactive materials – essential if an emergency response is to be planned – local authorities have little idea of what to expect. In practice, they are expected to rely on advice from the nuclear industry or central government about what action is needed.

Past experience suggests that public health might not always be a priority;

countermeasures might not be recommended. For minor incidents it is easy: in 1984, for example, Preston station was partially evacuated when a loose weather cover on a passing spent fuel flask raised suspicions of a possible leak. (The flask was checked and found safe.) Yet during Britain's worst nuclear accident to date – the 1957 Windscale fire – while site workers were told to keep windows closed and shelter indoors, nearby residents were neither warned nor officially informed. Preventing panic and bad publicity for the nuclear industry was seemingly more important than public health. As for the transport of spent fuel, the government's view was made clear in 1983: "Even in most extreme circumstances it is not envisaged that extensive evacuation would be necessary".[6]

Any decision to evacuate must, of course, weigh benefit against harm. For example, moving the residents of a sheltered housing scheme to avoid exposure of a few millisieverts may be more than offset by the anxiety caused, or more serious consequences for their health. However, other considerations show that decisions about evacuation have a political dimension. Whatever the advantages for public health, evacuation would incur economic penalties: industry and business would be disrupted; working hours would be lost; output, sales and profits could be threatened. If a release of radioactivity occurred in the centre of a large city, there would be every reason to avoid disrupting the system.

Not surprisingly, the nuclear industry has adopted, in effect, a cost/benefit approach to public health and contingency planning. This philosophy is illustrated in a review by the Health and Safety Executive (HSE) of a NAIR incident. It was not a transport event but an occasion when a cobalt-60 gamma source failed to retract into its shielded container while being used to radiograph a footbridge under construction in the centre of Newport bus depot, Monmouthshire. The occupants of a nearby hotel were evacuated and the area sealed off. The HSE listed the consequences of the incident: several thousand people inconvenienced; bus passengers denied access to the depot; buses delayed by re-routing; traffic inconvenienced by diversions; hotels, shops and offices closed for over half a day; 263 man-hours of police time consumed. The HSE questioned whether these "penalties" were necessary: "Where significant disruption to normal life would otherwise be occasioned and when this would cause a considerable financial penalty then a balanced judgement has to be made as to whether the detriment resulting from exposure is less important than the disruption to individuals and to the community".[7] Applied to a major accident in the heart of a city, it might well be judged that the economic disruption caused by evacuation would simply be too great a cost. After all, if a few extra cancers eventually appeared, many years after the event, would anyone trace the cause?

In practice, a decision to evacuate would be made in accordance with guidelines published by the NRPB. These are known as emergency reference levels (ERLs) and specify the "averted doses" which justify different counter-measures – sheltering, evacuation and (for reactor accidents) the administration of iodine tablets. They are intended mainly for an accident at a nuclear installation, but would also be applicable to transport accidents. In the case of evacuation, the ERL is an individual whole body dose of at least 30 mSv, incurred during the accident or from subsequent contamination.[8] (The dose might be received over several years.) Clearly evacuation would not be officially recommended just to avoid exposures of a few millisieverts. ERLs are striking evidence that official action to avoid radiation exposure – which may be up to thirty times the recommended annual dose – is not without its limits.

The nuclear juggernaut vs. local democracy

Local authority attitudes to emergency planning began to change during the 1980s. Conflicts with Whitehall over civil defence and a flood of disasters – including transport accidents like the Lockerbie air crash and the Kings Cross fire – reinforced a growing opinion that the role of local authorities in emergency and disaster planning needed to be redefined. Reassurances about civil nuclear power were also challenged, as were the assumptions underlying plans for dealing with accidents to radioactive materials in transit. The industry presumption that a major leak could never occur was simply not believed.

Many local authorities concluded that they needed to able to make their own assessment of the doses and dangers in the event of a radiation accident – an approach underlined by the experience of Chernobyl. As the radioactive cloud drifted across Europe, ignorance and confusion reigned. National governments reacted with conflicting advice on what levels of radiation were safe and what mitigating measures should be taken. A growing number, for example, have acquired their own monitoring equipment to give them independent, localised readings. In 1989, twenty of the London Boroughs backed the establishment of a radiation monitoring scheme for London. The scheme was set up not only to detect radiation from accidents at nuclear power stations (including those nearby in France) but also from transport incidents involving spent fuel or nuclear warheads.

Local and central government now have different and conflicting views about emergency planning. Central government assumes that if a local authority has no nuclear facility, it does not need to plan for nuclear accidents. The Home Office, in describing the Flask Emergency Plan, notes that "The likely scale of a flask mishap is such that no exceptional demands would be placed on local authorities emergency services".[9] Such a view is only tenable, however, if nuclear industry predictions are accepted without question. If they are not, then regardless of how remote it might be, there is always a chance that an incident far worse might occur. Common sense suggests that plans to respond to radiation accidents should be prepared for a full range of possibilities, including the worst conceivable accident.

Another response by local councils has been the declaration of "nuclear-free zones" (NFZs), some of which embody resolutions opposing the movement of radioactive materials. In Britain most NFZs – like the first, declared by Manchester City Council – have reflected concerns about nuclear weapons and civil defence:

> This Council, in the light of its pre-determined policy concerning the dangers of nuclear weapons, calls upon Her Majesty's Government to refrain from the manufacture or positioning of any nuclear weapons of any kind within the boundaries of our city.

Manchester's resolution, like those subsequently adopted by other councils, was a response to the political and military climate of the times, including the decision to allow Cruise missiles to be based in Britain. Not only did Cruise mobilise an upsurge of anti-nuclear opposition, but the missiles themselves were mobile; in the event of a crisis or war, they were to be driven up to 100 miles from their bases, thus making many local authority areas potential targets in the event of a war.

The transport of nuclear warheads became an increasingly important aspect of NFZ campaigning as local authorities discovered the existence of Mammoth Majors. The realisation that warhead convoys were travelling around Britain raised the possibility that a nuclear disaster could be caused not just by war or by an accident at a remote nuclear site. The country's nuclear defence policies could bring a radioactive catastrophe to communities otherwise miles away from nuclear bases or defence establishments like Aldermaston.

Local authority opposition also grew to the movement of other radioactive substances. The Greater London Council (GLC), for example, opposed the transport of nuclear materials for military use. Other councils have adopted NFZ policies critical of the civil nuclear industry as well, including the transport of irradiated fuel. South Yorkshire County Council was the first and, in March 1981, approved a series of recommendations specifically concerned with nuclear power and the transport of spent fuel, including:

- That demands be made of the Government to ensure that local emergency services and Emergency Planning Authorities should be made aware of the movement of irradiated fuel flasks and that British Rail and the CEGB be asked to make public the route and frequency of such movements.

- That the County Council consider that the present position in respect of the transport of irradiated nuclear fuel is unsatisfactory and that all such movements through the County should be stopped.

If the intention has been to raise the profile of nuclear and defence issues by information and publicity, then nuclear-free zones might be judged a success. However, NFZ declarations have obvious limitations. As broad statements of principle, they are often no more than gestures of defiance against the policies of central government. There is also an inconsistency in opposing the movement of nuclear warheads but not the radioactive materials which contribute to their production and deployment.

Some councils have therefore gone a step further and tried to implement specific policies on transport. Ipswich made one of the earliest attempts in 1980 when it tried to introduce a local by-law to ban the movement of spent fuel through its area. The town lies on the railway line from Sizewell nuclear power station to Sellafield and the proposed by-law reflected the council's concerns about the safety of spent fuel flasks if an accident occurred and their vulnerability to attack by terrorists. In the event, the council's efforts were unsuccessful; local authorities have no jurisdiction over land owned by British Rail.

The GLC tried an alternative approach around the same time. It too was concerned about accidents and terrorist attack and asked British Rail to reroute trains of spent fuel away from London. This would be entirely feasible for all spent fuel tavelling north from nuclear power stations in the south east, although it raises a moral dilemma: while fewer people would be affected by an accident if one occurred in a small town rather than a large city, the residents of, say, Reading or Ely would not thank the GLC if they had to risk what Londoners would not. In fact, the geography of Britain's railway network makes it impossible to plan routes from nuclear power stations to Sellafield without going through at least some densely populated areas. More to the point, if the nuclear industry accepted rerouting, it would be a tacit admission of potential danger. Perhaps for this reason alone, attempts to re-route flasks have been unsuccessful.

Other councils have tried to introduce a licensing system rather than an outright ban; although by refusing to grant a license the result would be the same. Having failed to persuade British Rail to re-route spent fuel away from London, the GLC approved a policy that "no (civil) nuclear materials should be transported through London without the specific authorisation of the GLC", a policy also adopted by several of the London Boroughs. Yet like Ipswich's attempt to introduce a by-law, the GLC had no powers to implement its policy.

Manchester City Council tried another tactic. In 1985 it drew up a local traffic regulation barring the movement of radioactive materials within certain areas. The highways department even designed an appropriate roadsign. Manchester's initiative stemmed from the city's inability to find out what radioactive materials were passing through the city. Apart from the wider aim of a "nuclear-free" Manchester, the initiative was designed to "flush out" information about radioactive shipments through the area. The hope was that, regardless of whether the regulation could actually be implemented, the act of proposing it would draw objections from the nuclear industry and other users of radioactive material who would then reveal what materials would be affected. But like others before it, the initiative floundered; traffic regulations require the approval of the Department of Transport which was not forthcoming.

The limitations of British local councils in the face of central government are more apparent when contrasted with the situation in some other countries. In Germany, the country's federal structure allows state authorities to exercise substantial control over the movement of radioactive materials. In 1988, for example, the export to Sweden of spent MOX fuel from the decommissioned Kahl nuclear power station was scuppered twice by state government opposition; after the authorities in Schleswig-Holstein withdrew a permit allowing the material to be transported through the port of Lübeck, a similar refusal by the government of Lower Saxony prevented its export through Emden.[10]

Transport issues were one of the factors behind the demise of the German reprocessing plant at Wackersdorf. Full use of the plant would have generated some 50 consignments of nuclear material each week, including irradiated fuel and plutonium. In 1985, the City of Nürnberg, about 50 kms west of Wackersdorf, commissioned a study by the "Gruppe Okologie" of the dangers of transporting radioactive materials to and from the proposed plant. The study assessed the consequences to the city of five "maximum possible accidents" and acts of sabotage: unlike the nuclear industry, it assumed that the containers and flasks to be used (including the Castor flask which brings imports to Sellafield) could be breached and release radioactivity in the city. The study concluded that large areas of Nürnberg would need to be evacuated in the event of a transport accident and cancer risks would significantly rise. The study's critical conclusions encouraged the city to take court action to stop the plant being completed. In 1989, after several years of sustained opposition – for many other reasons as well – the project was finally abandoned.

In the United States, over 200 local and state governments have passed transportation ordinances or bans since the mid-1970s. US local authorities have been able to stop the movement of radioactive materials, including irradiated fuel, to the considerable inconvenience of the industry. In 1990, for example, the radioactive containment vessel of the Shippingport nuclear reactor being decommissioned in Pennsylvania was transported to a waste disposal facility at Hanford in Washington by barge: an 8,100 mile journey via the Ohio and Mississippi Rivers, the Gulf of Mexico, the Panama Canal and up the Pacific coast – all to avoid travelling overland.[11]

The best-known example of local authority action has been New York City which, in the mid-1970s, introduced a system of licensing the movement of certain radioactive materials through its area. For almost ten years, the streets of Manhattan had provided a convenient route for road deliveries of spent fuel from a research reactor on Long Island. The Brookhaven National laboratory (BNL) sent flasks of irradiated highly enriched uranium (93 per cent) fuel from its High Flux Beam Reactor (HFBR) to the Savannah River reprocessing plant in South Carolina. Low level radioactive waste from BNL was dispatched to burial facilities. New York City was also traversed by road deliveries of plutonium travelling to and from JFK Airport. With the prospect of three nuclear power stations then under construction or being planned on Long Island (at Shoreham and Jamestown – all now cancelled or deferred) sending an extra 200 to 250 shipments of irradiated fuel through the streets of the city each year, the transport of radioactive materials became a controversial issue.

In 1976, the city's Board of Health unanimously passed an amendment to the New York City Health Code, effectively stopping the movement of all of such radioactive materials through the city. The adopted resolution meant that shipments of certain specified types of radioactive material were only allowed through the city if they had been authorised by a "Certificate of Emergency Transport", issued by the Department of Health. The specified materials included all but the smallest quantities of plutonium, uranium enriched above 25 per cent, and all types of spent fuel. The Resolution noted that a "Certificate will be issued for the most compelling reasons involving urgent public policy or national security interests transcending public health and safety concerns and that economic considerations alone will not be acceptable as justification for the issuance of such Certificate". In other words, the streets of New York were not to be used simply for the convenience of the nuclear industry because it was the shortest and cheapest route.

Not all radioactive materials were affected. The resolution stated that permission would be given for substances needed for certain medical uses such as teletherapy and heart pacemakers. Other radioactive materials intended for therapeutic radiology, biomedical research and educational purposes were exempted from control. However, by refusing to issue a certificate for the more dangerous types of material, substances like plutonium and spent fuel were effectively banned from New York. BNL, which prior to New York's ban had made 330 shipments of spent fuel through the city, responded by rerouting its radioactive materials through Connecticut. By using the ferry service across Long Island Sound to New

London, it was able to by-pass New York City, until similar restrictions were introduced in Connecticut.

New York's action was eventually overruled. The US Department of Transport (DOT) issued a new ruling entitled "Radioactive Materials, Routing and Driver Training Requirements" (otherwise known as HM-164) which stipulated that carriers of large quantities of radioactive material, such as spent fuel, must use "preferred routes" – either interstate highways or bypasses, or alternative highways designated by a State. The new rule – in effect, a national routing requirement – meant that spent fuel could be transported over any interstate highway. Despite an unsuccessful challenge in the federal courts, the DOT ruling overrode New York City's ban and on 1 January 1985, permission was given for shipments from Long Island to resume. Despite the setback, the City at least managed to negotiate several concessions with the Brookhaven Laboratory. BNL agreed to notify the City when shipments would occur, shipments were to be made only during the early hours of morning and on a previously selected route made known to the City, and both New York City and State would escort the shipment through their areas.

However, the fact that the ban remained in force for several years reflects a major difference between the US and British systems of government. Compared to the United States, where – as in Germany – the federal system devolves greater powers of administration to states and cities, local government in Britain can do little that does not need approval by or cannot be overturned by central government.

New York's limited success, and similar prohibitions enacted by other US local authorities which have remained in force, illustrate exactly why local councils in this country will never be allowed to have their way. By preventing Brookhaven National Laboratory from dispatching irradiated fuel, New York's action could eventually have closed the reactor. (It managed to continue working only by increasing the capacity of its spent fuel storage ponds.) If local authorities could control or stop the movement of radioactive materials, the nuclear industry would grind to a halt. The industry's vulnerability to disruption is recognised by the industry itself. Transport especially is crucial to its existence. A dramatic acknowledgement of this comes from an industry conference paper, one of whose authors is the managing director of NTL: ". . . the transport of radioactive materials . . . impinges on every section of the nuclear fuel cycle. Unless this common link works satisfactorily, the whole system would collapse . . ."[12]

18   The nuclear future?

At the beginning of the 1990s, the nuclear industry was in a state of uncertainty: while some of its operations were being closed or curtailed, others were planning to expand. Changes in both directions have implications for the transport of radioactive materials. On the military front, one change has already affected the industry; namely, the decision to reduce the size of the nuclear submarine fleet which in turn will reduce demand for nuclear fuel – hence the subsequent announcement of the closure of BNFL's military enrichment plant at Capenhurst.

On the civil side, the last years of the 1980s were marked by a succession of reactor closures. As the Magnox power stations approach the end of their lives, Berkeley and Hunterston "A" have already been shut down. In 1990, the UKAEA closed the SGHWR at Winfrith and the DIDO and PLUTO research reactors at Harwell – there are now no working reactors on that site. Dounreay's Prototype Fast Reactor may well cease operating in 1994 as a result of government cutbacks in funding. The most dramatic announcement was the decision to shelve plans to build three more PWRs after Sizewell "B", at least until after a review in 1994.

Reactor closures and the curtailment of nuclear generating capacity have a knock-on effect for companies supplying nuclear materials and fuel cycle services. BNFL has seen its domestic market retrench while the UKAEA has been forced to streamline its operations. For those who wish to see a nuclear-free Britain, cutbacks and closures are a welcome sign that common sense – if only of the economic kind – is finally beginning to prevail.

Yet such a verdict could be deceptive. After the Hinkley C public inquiry, planning permission was given for a third nuclear power station on that site; Hinkley C could well be the next new reactor if the 1994 review concludes that more nuclear power stations are needed. Even if it does not, further reviews may occur in years to come. In any case, BNFL are considering replacing Calder Hall and Chapelcross while for pro-nuclear die-hards, fast breeders continue to offer a panacea for long-term energy problems. In 1989, representatives from Britain, France and West Germany signed an agreement to develop a European Fast Reactor (EFR), for future use throughout Europe.

Moreover, if economic factors have undermined the industry's expansion in recent years, they may not always prevail in the future. Margaret Thatcher's view that "there are very powerful ecological, as well as strategic arguments for nuclear power"[1], is shared by many who believe that nuclear power will live to fight back. Strategic considerations especially have always been significant, including such unpredictable factors as the militancy of the coal mining unions and the supply of oil from overseas. A political battle over the privatisation of the coal industry or a crisis in the Middle East could well be the industry's saving grace.

It is also apparent that despite – or perhaps because of – the major upheavals in international relations, the military role of the nuclear industry will continue. Notwithstanding the demise of the Warsaw Pact, the government remains committed to a nuclear deterrent. As the Prime Minister said while the Warsaw Pact was disintegrating and within months of the Berlin Wall coming down: "Keeping Britain's nuclear deterrent has never been more important".[2] If necessary, new enemies will be found to justify its existence. Government thinking about nuclear proliferation has been voiced by Armed Forces Minister Archie Hamilton: "[Our service men], like me, would hate to face a future in which certain Third-world countries had nuclear weapons while Britain had none".[3]

In any case, the main political parties are committed to Britain's membership of NATO which is committed to the modernisation, if necessary, of its nuclear stockpile – no government in the immediate future is likely to denuclearise the Ministry of Defence. The nuclear submarine fleet will be maintained, albeit at a smaller size, and the nuclear weapons programme continued. Thus at the time of writing, the production of Trident warheads is well underway, with the prospect after that of a new generation of free-fall nuclear bombs to replace the aging WE177. Feasibility studies are underway for a new generation of nuclear submarines. The ongoing military activities of the nuclear industry will continue to justify a critical interest in the transport of its raw materials.

As for the civil side of the industry, the flagging home market has encouraged BNFL and the AEA to look overseas for new business. In 1990, BNFL announced the formation of a US subsidiary, BNFL Inc. to exploit marketing opportunities in the US for radioactive waste management and decommissioning·business. Later in the same year, International Nuclear Fuels Ltd (INFL) was set up – a major initiative by BNFL "to bring new business, access to new markets and increase BNFL's effective size and product range".[4] INFL consists of oxide fuel fabrication and reprocessing, hex production, enrichment and transport. It aims to "create strong long-term relationships with utilities and other companies with common interests and reduce dependence on the UK industry". Such objectives also underlie changes within the UKAEA which in 1989 launched nine new business sectors collectively known as AEA Technology. While this restructuring was primarily designed to win orders for R & D (in both nuclear and non-nuclear areas), the UKAEA has also sought new fuel cycle business to compensate for lost work in this area. Thus a weak domestic market is compensated for by selling fuel cycle services abroad. This has implications for transport: any decline in the movement of radioactive materials for the home market will be offset by more imports and exports.

Future nuclear trade

Foreign trade is becoming even more crucial to the prosperity of Britain's nuclear industry. Conversion, enrichment and reprocessing contracts with overseas customers will continue to provide valuable export earnings for the fuel cycle factories in Britain. At the "back end" of the fuel cycle, BNFL's THORP reprocessing plant illustrates how contracts with overseas customers are crucial to the company's prosperity; foreign trade not only accounts for most of its work, but also financed its construction. The economic viability of the plant depends upon reprocessing fuel from overseas, which in turn generates more shipments of radioactive material. Not only will separated uranium, plutonium and high level waste eventually be returned to the customer, but BNFL's efforts to drum up more reprocessing business for THORP into the twenty first century will ensure that "nuclear waste" from overseas continues to arrive in Britain.

Business prospects were boosted in 1989, when construction of the German reprocessing plant at Wackersdorf was abandoned. German utilities requiring reprocessing services are expected to send their spent fuel to Britain or France. Like North American sanctions against uranium imports from southern Africa, opposition in another country to the industry's activities has worked to BNFL's advantage.

In addition to the well-established fuel cycle operations of conversion, enrichment and reprocessing, Britain's nuclear industry has been developing new services to exploit further opportunities for overseas business. In 1984, BNFL inaugurated a new canning and assembly plant at Springfields for manufacturing 200 tonnes of PWR fuel per year. The decision to build it was a gamble that orders would be forthcoming from overseas customers. Although the new facility is supplying the initial load for Sizewell "B", its success depends upon exports.[5] BNFL's committment to PWR fuel business was underlined in 1989 when the company approved plans for a new oxide fuel complex including a plant to convert enriched hex to uranium dioxide powder and new facilities for manufacturing oxide fuels.[6]

For its part, the UKAEA announced in 1990 that "Dounreay is seeking to expand its overseas business in manufacturing and reprocessing MTR . . . fuel". New business being negotiated at the time included contracts to store and reprocess spent fuel from research reactors in Germany, Spain and the Netherlands. In addition to highly enriched uranium fuels, Dounreay is also aiming to manufacture low enriched uranium fuels for research reactors.[7]

New developments in the nuclear industry will result in new types of radioactive material being transported around Britain. One of AEA Technology's first projects has been a collaboration with BNFL to construct a demonstration MOX fuel plant on the AEA's site at Sellafield. The new facility will use plutonium from BNFL's adjoining reprocessing plant and be operational from 1993. The decision to build the pilot plant enabled BNFL to offer to supply MOX fuel for Japanese and Swiss customers. If sales look promising, BNFL hope to build a commercial-scale plant in the late 1990s.[8] German utilities are also potential customers; their recent reprocessing contracts with BNFL include options to have MOX fuel manufactured out of the recovered uranium and plutonium. BNFL have apparently concluded that the anti-nuclear climate in Germany which stopped Wakersdorf would also preclude the construction of a MOX plant in that country – another opportunity for BNFL to process and supply what other countries will not.

If the movement of radioactive materials has attracted attention in the past, there is every reason to suppose that it will continue to do so in the future. With substances like plutonium, whose transport has already proved controversial, any increase in the number of shipments will be accompanied by an increase in public concern. But it will not just be a rerun of the past; there will be new materials moving around. Shipments of high level waste and MOX fuel – and eventually irradiated MOX fuel – are likely to attract attention. If the European Demonstration Reprocessing Plant is ever built at Dounreay, there would also be imports of irradiated fuel from European fast reactors travelling the length of Britain. Such new materials would bring new potential hazards. (For example, neutron emissions from irradiated MOX fuel are some 15 times higher than ordinary power station fuels.[9])

New transport routes may open up. BNFL have reportedly been pursuing business in the former Eastern Bloc countries, offering to reprocess irradiated fuel from Czechoslovakia[10] and the former German Democratic Republic.[11] The company have also been eyeing the American market with several US utilities expressing interest in Sellafield's reprocessing services.[12]

While health and safety concerns will undoubtedly remain a priority, the industry's efforts to secure more overseas business will also raise wider issues. The problems of waste disposal, the spectre of terrorism and weapons proliferation, the inevitable shroud of secrecy, will all impinge on the transport of radioactive materials – regardless of how safe it might be.

Many people remain unconvinced that the transport of radioactive materials is safe. In particular, the possibility of a major accident releasing radioactivity continues to generate concern. On a technical level there are clearly actions which could be taken by the industry to reassure its critics.

The criteria for testing transport containers could be made more rigorous. As previous chapters have noted, the fire and impact tests recommended by the IAEA have been criticised several times; the fact that more severe criteria have been suggested or adopted within the industry itself further undermines confidence in IAEA standards. For example, when Ontario Hydro carried out a fire test for a new flask for Canadian spent fuel they doubled the duration of the fire, performing a one hour test instead of the 30 minutes recommended by the IAEA – to the chagrin of others in the industry.[13] If more rigorous tests are justified for one type of flask, should they not also be applied to others?

Changes have been suggested to the industry's plans for responding to an emergency. As a report prepared for the Nuclear Free Zones National Steering Committee noted: "It is doubtful whether the public would be reassured by the fact that the very company responsible for contaminating their locality are also responsible for clearing up the mess and subsequent monitoring".[14] It is interesting to compare the long-standing NAIR scheme with NIREP: while the former is co-ordinated by the NRPB – an organisation which likes to assert its independence from the production side of the industry – the latter is organised by the UKAEA constabulary whose interests are usually directed towards the maintenance of secrecy; indeed, some would regard them as symbols of a "nuclear state". If nothing else, a more generous supply of information would undoubtedly please local authorities and the emergency services.

Secrecy is associated with a general lack of accountability. Parliamentary control over the nuclear industry is largely theoretical. Organisations like BNFL and the UKAEA are answerable only to Ministers and their government departments. The nuclear industry can ignore the recommendations of parliamentary select committees while members of parliament have at best only an indirect influence over the industry's activities. The 1986 report of the Select Committee on the Environment provides a good example. The all-party committee was concerned principally with waste disposal and reprocessing, but also looked at some aspects of the transport of radioactive materials. One was the possibility that a train carrying spent fuel might collide in a tunnel with another transporting inflammable material. (The committee was taking evidence shortly after the Summit Tunnel fire.) The committee concluded:

> We therefore recommend that British Rail should re-consider the possibility of timetabling movements of radioactive material so that the remote chance of an accident involving inflammable materials in a tunnel can be made even more remote.[15]

In the next paragraph, the committee expressed its concern about the transport of hazardous radioactive materials by air:

> We recommend that wherever possible, radioactive waste, especially spent fuel, high-level waste and plutonium should be carried by rail in preference to all other modes of transport. The carriage by air of all except the very lowest levels of radioactive materials should be prohibited.

However, while British Rail has reportedly reorganised its timetables to heed the committee's advice[16], as far as plutonium is concerned, the nuclear industry has ignored it. Although future exports of plutonium to Japan have switched from air to sea, this was due entirely to the difficulties of meeting the stringent US criteria for testing plutonium containers – and in any case deliveries of plutonium to Europe continue to travel by air.

Yet technical and administrative issues are clearly not the whole story. Any attempt to explain the disparate concerns over transport must acknowledge that there is far more involved than the relatively minor routine hazards of normal transport or even the potential for a major disaster. Nowhere is this more evident than in the enduring opposition to the movement of spent fuel. Whatever the hazards – real, potential or imagined – spent fuel transport is also related to reprocessing and therefore to the issues associated with that operation: the extraction of large quantities of plutonium, the potential dangers of a "plutonium economy", weapons proliferation, the discharge of radioactivity to the environment, the disposal of radioactive waste, and the health hazards to the surrounding community. For example, the Irish Sea now contains – on the government's own admission – radioactive isotopes of the following elements: americium, antimony, argon, caesium, carbon, cerium, cobalt, curium, europium, hydrogen (in the form of tritium), iodine, krypton, manganese, niobium, plutonium, ruthenium, silver, strontium, technetium, zinc and zirconium – an appalling cocktail of pollution from Sellafield and largely derived from its reprocessing operations.[17] (Some discharges are the result of Sellafield's military activities.) If the nuclear industry despairs of the attention paid to the movement of spent fuel flasks – whose contents are largely responsible for that pollution – they have clearly failed to appreciate transport's wider significance.

For this reason, the transport of the industry's raw materials has acquired a symbolic and tactical importance: the movement of radioactive materials presents an ideal opportunity for campaigning. Nuclear sites cannot function without a supply of radioactive materials. For this reason, the transport network has been viewed as the industry's umbilical cord: sever this link and the production of nuclear weapons and nuclear elecricity – and their radiation and waste – would eventually grind to a halt. To stretch the biological analogy further, the supply of radioactive material is the nuclear industry's life-blood, without which the patient would die.

Even more important is a fundamental clash of values. It is impossible to explain any controversy about nuclear issues without recognising that the industry and its critics approach the subject with different preconceptions; issues like health and safety are judged by different criteria. The industry views its critics as irrational, emotional and opposed to science. Critics view the industry as secretive, arrogant and above all, inhuman. As Madame Curie used to say: "In science we must be interested in things, not in persons".[18] Differences often arise because many of the industry's assumptions rest upon value judgements. An industry opinion from a conference paper on the risks of transporting plutonium nitrate illustrates the point:

> An assessment was also made of the consequences of losing a package at sea. From this assessment it was concluded that neither consumption of sea-food contaminated by any plutonium released, or exposure resulting from plutonium being washed ashore and contaminating the shore line constituted an unacceptable level of risk.[19]

Such conclusions are ultimately subjective. What is acceptable inside the nuclear industry is not necessarily acceptable outside. The amount of plutonium released by such an accident may be minimal; but if the public decides that it wants to eat sea-food totally free of plutonium, that is the public's right – regardless of whether the risks associated with contamination are large or small.

If the industry's actions had no effect on the rest of the world, such technicalities might not matter. But unfortunately the rest of the world *is* affected. Every living being on the planet contains fallout from the testing of nuclear weapons. Many have absorbed radiation from accidents like Chernobyl. Local communities play host – willingly or not – to radioactive materials on their way from one nuclear site to another. More importantly, people could be affected if something went wrong.

Notes and references

Chapter 1

1 O'Sullivan, R.A., "International consensus for the safe transport of radioactive material: An experience to imitate", IAEA Bulletin, 3/1988, p.32
2 *The Transport of Civil Plutonium by Air*, Advisory Committee on the Safe Transport of Radioactive Materials, 1988, paragraph 5.1.2, p.12

Chapter 2

1 *Symposium on the Transport of Radioactive Materials*, UKAEA, Harwell, 1963, p.8
2 *Safe transport of radioactive material*, IAEA booklet, Vienna, 1985
3 *Symposium on the Transport of Radioactive Materials*, UKAEA, Harwell, 1963, p.14

Chapter 3

1 Quoted in *Nuclear Europe*, 7-8/1989, p.65
2 *Rossing Fact Book*, Rossing Uranium Ltd. Windhoek, 1989
3 *Nuclear Engineering International*, August 1990, p.18
4 *ATOM*, April 1990, p.29
5 *Financial Times*, 5 November 1985
6 *Nuclear Fuel*, 12 December 1988
7 Roberts, Alun, "Stolen Uranium: Labour breaks a promise" in *New Statesman*, 2 June 1978
8 *Financial Times*, 30 March 1983
9 *The Guardian*, 21 December 1984
10 Letter from Malcolm Rifkind to the Namibia Support Committee, 27 November 1984
11 15,000 tonnes: *Hansard (Commons)*, 11 May 1988, col. 112w
12 Statistics derived from "World List of Nuclear Power Plants" *Nuclear News*, February 1991
13 *Energy in Europe*, 12/1989 p.42
14 *Hansard (Commons)* 19 May 1989, col. 314w

15 Page, H., (BNFL) "UK Experience of Production of Uranium Fluorides", conference paper from *Production of Yellowcake and Uranium Fluorides* (Proceedings of an Advisory Group meeting), IAEA, Vienna, 1980

16 *New Statesman*, 31 July 1987, p.6

17 *Hansard (Commons)*, 19 April 1988, col. 382w

18 Clark, David, (CANUC) "Namibia's Nuclear Nightmare", in SCRAM Journal, Jan/Feb 1988, pp.12/13

19 Stephany, Manfred, "Influence of stockpiles on the market for natural uranium" in *Uranium and Nuclear Energy: 1981*, Butterworth Scientific Ltd, 1982

20 Sizewell Inquiry Day 274, p.66 F

21 Castle, Barbara, *The Castle Diaries 1974-76*, Weidenfeld & Nicolson. London, 1980, p.635

22 Text of evidence given by the Rt. Hon. Tony Benn MP to the commission of inquiry into the legal aspects of Rio Tinto Zinc's mining operations in southern Africa, Bristol cathedral, 27 November 1982

23 *The Guardian*, 2 August 1988

24 *Hansard (Lords)*, 30 November 1988, col. 300 (oral answers)

25 *Hansard (Lords)*, 30 November 1988, col 300. (oral answers)

26 *Hansard (Commons)*, 19 May 1989, col 314w

Chapter 4

1 *ATOM*, November 1984, p.13

2 The Health Physics Manager's comments were made to the local liaison committee. Minutes of the Springfields Liaison Committee meeting, 7 December 1984

3 *ATOM*, November 1984, p.13

4 "Beach scare", *Daily Mail*, 16 February 1985.

5 *Lloyd's List*, 16 July 1985

6 *Nuclear Fuel*, 12 August 1985, p.14

7 *Hansard (Commons)*, 30 July 1982, cols. 866/867w

8 *Nuclear Fuel*, 25 June 1990

9 *Nuexco Monthly Review* #202, June 1985, p.29

10 *Hansard (Commons)*, 6 December 1988, col. 103w

11 Nuclear Europe, 11-12/1989, p.51

12 *BNFL Annual report* 1985/86, p.18 & A.R. 1988/89 p.14

13 *ATOM*, April 1991, p.3

14 Spoor, N.L., and N.T. Harrison, *Emergency exposure levels for natural uranium* (NRPB-R111, NRPB, Didcot, 1980, p.4

15 Quoted in "Lorry overturns its load of nuclear fuel in by-pass crash", *The Times*, 20 May 1977

16 *The Waste Paper*, Sierra Club Atlantic Chapter Radioactive Waste Campaign, Spring 1982, p.1

17 *An Assessment of the Risk of Transporting Uranium Hexafluoride by Truck and Train* (PNL-2211), August 1978, Batelle

18 *Nucleonics Week*, 7 July 1977, p.10

19 *Nuclear Fuel*, 30 June 1986, p.7

20 *Nuclear Fuel*, 30 June 1986, p.7

21 "UF6 cylinders pulled from service because of cracks in small valves", *Nuclear Fuel*, 24 August 1987, p.11

Chapter 5

1 *Hansard (Commons)*, 25 July 1990, col. 471 (Options for change)
2 *Hansard (Commons)*, 9 February 1990, col. 815w
3 *Hansard (Commons)*, 22 April 1959 col. 393, Nuclear Fuel Elements (Manufacture)
4 *The Energy Daily*, 16 January 1979. p.2
5 UKAEA 9th. Annual report (1962-63), para. 92, p.18; *Hansard (Commons)*, 27 January 1982, col. 384w; *Hansard (Commons)*, 9 May 1988, col. 2w
6 According to US Admiral McKee, enrichment for naval reactors is "about 97%" (Testimony of US Admiral McKee before the Procurement and Military Nuclear Systems Subcommittee of the House of Representatives Committee on Armed Services, Hearing on H.R. 4526, Department of Energy National Security Programs Authorization Act for Fiscal Years 1987 and 1988, 20 February 1986, p.26.) Although this refers to US naval fuel, it can be assumed that British nuclear submarine reactors, whose design was derived from the US model supplied for *Dreadnought*, use uranium enriched to the same level
7 The Guardian, 2 May 1986, p.4
8 6th Report from the Defence Committee 1984-85 "The Trident Programme", p.9
9 6th Report from the Defence Committee 1984-85 "The Trident Programme", p.xxii. (seven years); *Hansard (Commons)*, 3 July 1985, col. 172w (seven or eight years)
10 Rolls-Royce letter to Councillor Lennox, Public Protection Committee, Derbyshire County Council, 7 August 1984
11 *Hansard (Commons)*, 27 January 1982, col. 384w.
12 1st Report from the Select Committee on Energy (Session 1980–81), *The Government's statement on the new nuclear power programme*, Vol. III, p897 (Supplementary Memorandum by RR&A Ltd)
13 Rolls-Royce letter to Councillor Lennox, Public Protection Committee, Derbyshire County Council, 7 August 1984
14 *Chatham, Rochester & Gillingham News*, 15 April 1977
15 E.g. *New Statesman*, 10 April 1981, p.6
16 E.g. *Chatham, Rochester & Gillingham News*, 15 April 1977 (in an article headlined "Nuclear scaremongers who trade on ignorance"!)
17 *Hansard (Commons)*, 5 March 1984, cols. 472/3w
18 *The Scotsman*, 28 October 1981
19 J. Simpson, *The Independent Nuclear State*, Macmillan, London. 1986, Appendix 7
20 *Nuclear Fuel*, 13.4.81, p.9
21 J. Simpson, *The Independent Nuclear State*, Macmillan, London. 1986, p.201
22 *Nuclear Fuel*, 3.11.86
23 *Nuclear Fuel*, 11.7.88, p.9
24 *Nuclear Fuel*, 31.10.88, p.13
25 *Nuclear Engineering International*, June 1989, p.72

26 quoted in *The Glasgow Herald*, 10.1.91
27 *ATOM*, March 1991, p.3; *Scram Journal*, April/May 1991 p.6. For details of the road route taken to Dounreay see *Hansard (Commons)* 28 March, 1991, cols. 506/7w
28 Yemel'yanov, V.S., and A.I. Yevstyukhin, *The Metallurgy of Nuclear Fuel*, Pergamon Press, Oxford, 1969, p.444
29 Bellamy, R.G., and N.A. Hill (Harwell), *Extraction and Metallurgy of Uranium, Thorium and Beryllium*, The Macmillan Company, New York, 1963, p.186
30 Coleby, D. "Forum: Shipment of Spent Fuel from Australia", *Nuclear Spectrum*, Australian Atomic Energy Commission, p.9

Chapter 6

1 "Sizewell N-waste train halted by protesters", *East Anglian Daily Times*, 13 August 1982, front page
2 These figures are taken from a Nuclear Electric plc "packing sheet" (#HPB000624) for a consignment of AGR fuel dispatched from Hinkley Point B in flask #E41A to Sellafield on 3 January 1991
3 Hansard (Commons), 3 March 1980, col. 12w; ATOM, August 1963, pp.240–1
4 *Torness Power Station*, Precognition by E.G. Watt on "Operational and safety aspects of the proposed railhead" (SSEB evidence to the Skateraw Railhead Public Inquiry), October 1984, Appendix 3
5 *Hansard (Commons)*, 24 April 1990 col. 143w
6 Flowers, B., *Nuclear Power and the Environment*, the sixth report of the Royal Commission on Environmental Pollution, Cmnd. 6618, HMSO, 1976, Chapter 8, paragraph 421
7 *Hansard (Commons)*, 30 April 1990
8 Flowers, B., *Nuclear Power and the Environment*, the sixth report of the Royal Commission on Environmental Pollution, Cmnd. 6618, HMSO, 1976, Chapter 8, paragraph 420
9 *Carrying the Can*, The Ecology Party (North London Branch), 1980, p.43
10 *Railnews*, February 1985, p.5
11 *The Guardian*, 4 March 1983, p.3
12 *Transport of Irradiated Nuclear Fuel*, CEGB leaflet, August 1983
13 Transcript of the Channel Four Television programme, *Waste Not? Want Not?* (first transmitted 5 December 1985), p.11
14 *Nuclear Engineering International*, September 1989 p.42
15 *Nuclear Fuel*, 5 May 1986, pp. 6/7; see also: "British utilities to go ahead with dry storage for spent fuel", *Nucleonics Week*, 22 September 1988, p.8
16 *Nuclear Fuel*, 24 June 1991, p.17
17 *Nuclear Engineering International*, October 1990, p.11

Chapter 7

1   *Hansard (Commons)*, 1 February 1988, col 472w
2   *Nuclear Engineering International*, June 1986, p.12
3   Jean Emery "Victims of Radiation", in *The Ecologist*, Vol 16 No.4/5, 1986, p.191
4   *Hansard (Commons)*, 17 January 1979, cols. 747/8w; *Nuclear Fuel*, 13 April 1981, p.8
5   *Hansard (Commons)*, 17 January 1979, cols. 748w
6   *NTL Data Booklet 1973 – 1985*
7   Parliamentary reply from David Mitchell (Department of Transport) to Dafydd Elis Thomas MP, 12 December 1986
8   Harding, C.G.F., "A welcoming approach to winning support", *Uranium and Nuclear Energy: 1987*, Uranium Institute, London, 1988, p.27
9   *Hansard (Commons)*, 14 November 1988, col. 426w
10 *The Observer*, 2.9.90 p.1
11 *Transport of spent nuclear fuel*, BNFL leaflet, 1984, p.6
12 "Ill-fated "Gateway" was unseaworthy", *Lloyd's List*, 6 March 1984, p.1
13 *Lloyd's List*, 2 May 1987, p.1
14 *Hansard (Commons)*, 14 June 1988, col 128w
15 *Nucleonics Week*, 23 August 1990, p.6
16 *Transport of spent nuclear fuel*, BNFL leaflet, 1984, p.9
17 report by Sir Frank Layfield, Sizewell B public inquiry, Summary of Conclusions and Recommendations, 1987, p.22
18 Resnikoff, Marvin, *The Next Nuclear Gamble*, Council on Economic Priorities, New York, 1983, p.211
19 Ross, Marc, "The transportation of radioactive wastes: the possibility of release of cesium" in, *The Nuclear Fuel Cycle*, The Union of Concerned Scientists, The MIT Press, Cambridge, Massachusetts, 1975

Chapter 8

1   Gowing, Margaret, *Independence and Deterrence: Britain and Atomic Energy 1945 – 1952*, Vol 2, Macmillan, London 1974, pp. 241/2
2   Flowers, B., (Chair), *Nuclear Power and the Environment*, 6th report of the Royal Commission on Environmental Pollution, Cmnd 6618, 1976. Recommendation #44
3   Tunaboylu, K., W. Hunzinger, U. Tillessen, *Risks of Plutonium Transportation*, Proceedings of a conference on the Packaging and Transport of Radioactive Materials (PATRAM) 1986, IAEA, Vienna, 1986, p.328
4   *Hansard (Commons)*, 9 March 1988, cols. 205/6w
5   *Hansard (Commons)*, 13 March 1980, col. 684w
6   J. Chicken, *Summary of the risk assessment made of the transport of plutonium nitrate*, PATRAM 1980, IAEA, Vienna, 1980, p.913
7   *ATOM*, January 1977, p.4
8   *Daily Express*, 17 & 18 May 1977
9   *The Times*, 17 January 1978

10 Dounreay Inquiry, Day 18, 1986, p.86
11 Dounreay Inquiry, Day 18, 1986, p.83 E
12 *Hansard (Commons)*, 19 November 1962, col 974 (Atomic Energy Establishment, Capenhurst)
13 *Hansard (Commons)*, 26 January 1982, col. 325w
14 quoted in Stott and Taylor, *The Nuclear Controversy*, Town and Country Planning Association, London 1980, p.101
15 Fifth Report from the Select Committee on Estimates, Session 1958–59, *United Kingdom Atomic Energy Authority*, p.216
16 Fishlock, D., "Aldermaston: 'enough plutonium to buy a frigate'", article in *Financial Times*, 24 May 1985
17 Department of Energy News Release, 5 December 1989
18 Department of Energy News Release, 18 October 1990
19 Simpson, J., *The Independent Nuclear State*, Macmillan, 1986, Appendix 7
20 *Hansard (Commons)*: 20.7.90 col.760w
21 *Hansard (Commons)*, 14 May 1981, col. 337w
22 *Hansard (Commons)*, 18 February 1987, col 633w
23 *Near-Term Plutonium Market Outlook* (ORNL/Sub/83-40111/1), report prepared for the US Department of Energy by the Nuclear Assurance Corporation, Norcross, Georgia, March 1983, p.12
24 *NUKE INFO TOKYO*, Mar/Apr 1990 No 16, p.3
25 *Hansard (Commons)*, 6 November 1986, col. 547w; *Hansard (Commons)*, 12 December 1986, col. 240w
26 Fishlock, D., "US plutonium for fast breeder tests" in *Financial Times*, 7 November 1979
27 Salmon, A., *The Transport of Plutonium by Air*, BNFL report SG1(87)36, December 1987
28 *Nuclear Engineering International*, March 1991, p.12
29 Salmon, A., op cit
30 Dounreay Inquiry, Day 18, 1986, p.79 E
31 Salmon, A., op cit
32 *Hansard (Commons)*, 22 December 1988, cols 411/2w
33 Solon, Dr Leonard, "Public Health Aspects of Transportation of Radioactive Materials in Large Urban Areas", from proceedings of the 8th. National Conference on Radiation Control 1976–7, published by US Department of Health, Education and Welfare, Public Health Service, Food and Drug Administration, Bureau of Radiological Health, Rockville, Maryland.
34 Dounreay Inquiry Day 18, 1986 p.81
35 Wilson, P.W., "Packaging for the Transport of Plutonium", PATRAM 1986, Vienna, 1986, p.77
36 Dounreay Inquiry, Day 18, 1986, p.93 B
37 Dounreay Inquiry Day 18, 1986, p.92 G (evidence of Mr. P.W. Wilson)
38 Brown, M.L., et al, (UKAEA) *Specification of test criteria for containers to be used in the air transport of plutonium (Nuclear science & technology series)*, Commission of the European Communities, Luxembourg, 1980. paragraph 2.5.2
39 Ibid. paragraph 8.4
40 Ibid. paragraph 10
41 Dounreay Inquiry Day 18, 1986, p.89 C (evidence of Mr. P.W. Wilson)

42 Brown, M.L., et al, op cit. paragraph 1.2
43 *Daily Telegraph*, 24 January 1979. p.5

Chapter 9

1   Defence Estimates 1987 p.39 and 41
2   Defence Estimates 1987 p.41
3   *Hansard (Commons)*, 23 February 1988, cols. 133/134 (oral answers)
4   Letter from G.G. Barlow (MoD) to the National Steering Committee Nuclear Free Local Authorities, 22 October 1990
5   *Hansard (Commons)*, 15 July 1987, col. 1129 and *Independent*, 15 July 1987
6   "Convoy crash may be quizzed", *Lennox Herald*, 28 June 1985
7   *Hansard (Commons)*, 26 November 1985, col. 741w
8   "The Progress of the Trident Programme", 8th. Report of the Defence Committee, 1990–91, A47
9   *Hansard (Commons)*, 30 April 1985, col. 120w
10 "Staff shortages are main constraint on production", *Independent*, 26.1.88, p.2
11 *Times*, 25 February 1966
12 "East Anglia's nuclear escape", *Guardian*, 6 November 1979, p.21
13 *Hansard (Commons)*, 5 December 1957, col. 608 (oral answers)
14 Blakeway, Denys and Sue Lloyd-Roberts, *Fields of Thunder*, Counterpoint, Unwin Paperbacks, London, 1985, p.140.
15 *Hansard (Commons)*, 22.11.82, col. 361w
16 *Preparedness for Nuclear Accidents* (GAO/NSIAD-87-15), Appendix 5 "Nuclear Weapon Accident Risks to Public Health and Safety"
17 "B-52 crash nearly triggered N-blast", *Guardian*, 15 September 1983 p.6
18 *Nuclear Weapons Safety*, report of the Panel on Nuclear Weapons Safety of the Committee on Armed Services, House of Representatives, 101 Congress, 2nd. Session. USGPO, Washington, December 1990
19 *Hansard (Commons)*, 14 June 1990, col. 325w
20 *Hansard (Commons)*, 14 November 1988, col 415w
21 *Nuclear Weapons Safety*, report of the Panel on Nuclear Weapons Safety of the Committee on Armed Services, House of Representatives, 101 Congress, 2nd. Session. USGPO, Washington, December 1990
22 *Preparedness for Nuclear Accidents* (GAO/NSIAD-87-15), Appendix 5 "Nuclear Weapon Accident Risks to Public Health and Safety"
23 Jackson Davies, W., *Nuclear accident aboard a naval vessel homeported at Staten Island, New York. Quantitive analysis of a hypothetical accident scenario*, quoted in: *Safety of British Nuclear Weapon Designs*, BASIC Report 91.2
24 *Hansard (Commons)*, 10 December 1957, cols. 1072/3 (oral answers)
25 "Atomic dust used in training", *Independent*, 15 July 1987; "Radioactive material still used in training" *Independent*, 16 July 1987
26 *Hansard (Commons)*, 15.7.87, col. 1135 Military Exercises (Radioactive Contamination)
27 Letter from B.R. Mann (MoD) to Dr. P.R. Webber, Deputy Director of Emergency Planning, South Yorkshire Fire and Civil Defence Authority, 28 August 1987

28 *Hansard (Commons)*, 15.7.87, col. 1130 Military Exercises (Radioactive Contamination)
29 Helliwell & Spink, *The Transport of Plutonium by Air and Sea*, European Proliferation Information Centre, London, 1989, pp.92–94
30 "Complaint by bomb town", *Times*, 24 May 1968, p.6

Chapter 10

1 Farkas, Joseph, *Irradiation of Dry Food Ingredients*, CRC Press Inc. Boca Raton, Florida, 1988
2 The 1st Report from the Select Committee on Energy, Session 1980–81: *The Government's Statement on the New Nuclear Power Programme*, Vol 3 (Supplementary Memorandum by Rolls Royce & Associates), pp. 895 & 897
3 Shaw, K.B., J.S. Hughes, T.D, Goodimng and L. McDonough: *Review of the Radiological Consequences Resulting from Accidents and Incidents Involving the Transport of Radioactive Materials in the UK from 1964 to 1988*, NRPB, Chilton 1990, p.3
4 Ibid
5 Gelder, R., *Radiological Impact of the Normal Transport of Radioactive Materials by Air*, NRPB, Chilton 1990, p.2
6 Gelder, R., op cit, p.3
7 Gelder, R., J.H. Mairs, K.B. Shaw "Assessments of radiation exposures from two transport operations in the United Kingdom", reprint from *Packaging and Transportation of Radioactive Materials (PATRAM '86)*, IAEA, Vienna, 1987, p.354
8 From evidence by Troy Wade (Assistant Secretary, Defense Programmes, DOE) to Hearings being held before the Sub-Committee on Energy & Water Development Appropriations of the Home Appropriations Committee. 7 March 1989, p.805
9 *Nuclear Fuel*, 10 July 1989
10 *Nuclear Fuel*, 10 July 1989
11 *Nucleonics Week*, 27 February, 1986 pp.14/15
12 *Hansard (Commons)*, 15.2.89 col. 271/272w
13 *Hansard (Commons)*, 27 June 1989, col. 389w
14 For example, General Electric's IF-300 spent fuel flask (for use in the United States) and the JSU-1 model spent fuel flask being developed in Japan. In Britain, depleted uranium shielding is incorporated in an American-designed technetium generator imported for use in this country.
15 BNFL letter of 6 February 1987 to SCRAM (Scottish Campaign to Resist the Atomic Menace)
16 CEGB letter to author, 15.1.87
17 Sizewell B Power Station Public Inquiry, CEGB Statement of Case, Vol 2, paragraph 28.1, p.178
18 "CEGB's operational experience in the transport of radioactive material packages", *Nuclear Energy*, October 1990, p.354

Chapter 11

1 *Hansard (Commons)*, 13 July 1989, col. 634w
2 Gowing, M., *Independence and Deterrence: Britain and Atomic Energy 1945 – 1952*, Macmillan, London Vol 2 p450
3 *Hansard (Commons)*, 25 February 1986, col. 817
4 Department of the Environment, *Statistical Bulletin (90)3: Radioactivity*, Government Statistical Service, July 1990, p.13
5 Ibid
6 *Power News*, June 1980, p.4
7 *First Report of the Environment Select Committee 1985/86., p.165*
8 Mummery G.B., and R.F. Pannett, "CEGB's operational experience in the transport of radioactive material packages", in *Nuclear Energy*, 1990, 29 No. 5, October, p.355
9 *New York Times*, 21 May 1976
10 *Hansard (Commons)*, 19 June 1990, col 788 (oral answers)
11 Nirex, *The Way Forward*, undated discussion document, p.31
12 *Hansard (Commons)*, 15 February 1989, col. 270w
13 *ATOM* July 1988, p.41
14 *Hansard (Commons)*, 5 June 1990, col. 410w
15 Nuclear Engineering International, September 1989, p.39
16 *Hansard (Commons)*, 2 May 1986, cols. 502/3w
17 Mairs, J.H., D.J.Bigley, K.B.Shaw, *The radiological impact associated with the options available for the transportation of Intermediate Level Wastes with respect to a Minimum Power Scenario*, NRPB, Chilton, December 1984. NRPB Ref: 7910 2348. p.1
18 *ATOM*, November 1987, p.9
19 Flowers Report, p.137
20 *Hansard (Commons)*, 20 May 1988, col. 604w
21 Transcript of the Channel 4 programme "Waste Not Want Not", broadcast in December 1985
22 *ATOM*, November 1984, p.14

Chapter 12

1 "Nuclear vehicle stranded by rush-hour breakdown", *Newbury Weekly News*, 25 July 1985, p.1
2 A. Lovins, "Nuclear weapons and power-reactor plutonium", *Nature*, 28 February 1980, p.817
3 *Nuclear Fuel*, 12 November 1990, p.8
4 Barham, K.W.J., D. Hart, J. Nelson and R.A. Stevens "Production and destination of British civil plutonium": article in *Nature*, Vol 317, 19 September 1985, p.217
5 For example: *Hansard (Commons)*, 3 December 1984, col. 24w
6 "US Officials Admit Use of Plutonium From Britain to Make Nuclear Arms", *International Herald Tribune*, 19 March 1984. p.1
7 Ibid

8   See Barnham, K., "Calculating the plutonium in spent fuel elements", in *Plutonium and Security – The Military Aspects of the Plutonium Economy*, ed. F. Barnaby, Macmillan, London, 1991

9   *Hansard (Commons)*, 24 July 1981, col.255w

10 *Hansard (Commons)*, 23 January 1985, col.425w

11 *Guardian*, 25 April 1985

12 *Guardian*, 25 April 1985

13 *Hansard (Commons)*, 15 April 1986, col.330w

14 *Hansard (Commons)*, 18 April 1986, col.525w

15 Evidence of the UKAEA (Production Group and Development and Engineering Group) published in the *Fifth Report from the Select Committee on Estimates*, Session 1958-59, H.M.S.O. London 1959. p. xxxix

16 Malone, Peter, *The British Nuclear Deterrent*, Croom and Helm, London, 1984 p.63

17 *Hansard (Commons)*, 9 March 1983, col 403w

18 Pugh, O., "Fuel Cycle Operations at Dounreay": in *The Nuclear Engineer*, Vol 26 No. 4, July/August 1985

19 *Nuclear Research Reactors in the World*, IAEA, Vienna, May 1986. p.71. See also *Hansard (Commons)*, 5 March, 1984 cols. 472/473w.

20 *Nuclear Fuel*, 14 January 1985, pp.12/13

21 According to the report in Nuclear Fuel, 14 January 1985, pp.12/13, the United Kingdom was one of at least seven countries which had returned HEU fuel to the US before 1983

22 Johnson, K.D.B., "A History of the UK Nuclear Fuel Cycle": in *The Nuclear Engineer*, Vol 25 No. 3, May/June 1984

23 *Times*, 28 April 1976

24 *UKAEA 17th Annual Report* 1971-72, p.113

25 *UKAEA 17th Annual Report* 1971-72, p.113

26 *Hansard (Commons)*, 8 March 1979, cols. 777/778w

27 *Hansard (Commons)*, 8 March 1979, col. 777w

28 *Hansard (Commons)*, 17 January 1985, col. 184w

29 *Hansard (Commons)*, 17 January 1985, col. 184w

30 *Hansard (Commons)*, 17 January 1985, col. 184w

31 *Guardian*, 11 May 1989, p.4

32 *Preston Leader*, 29.8.85

33 *Hansard (Commons)*, 17 December 1984

34 *Preston Mail*, 19.9.85

35 Minutes of the 1st Report of the Environment Select Committee 1985–86, para 515

36 *Nucleonics Week*, 10 November, 1988, p.1. (quoting BNFL)

37 *Hansard (Commons)*, 9 March, 1983, col. 405w

38 *Hansard (Commons)*, 14 November 1988, col. 425w

39 Fishlock, D., "Isotope enrichment on the boil", *Financial Times*, 6 January 1986

40 *UKAEA 9th Annual Report*, paragraph 240. UKAEA 11th Annual report, paragraph 14

41 *Times*, 21 February 1963, p.8

42 EdF article by M. Lammers quoted in *Sunday Times*, 14 August 1983, and in part in *ATOM*, March 1986 p.45 (quoting *Hansard*)

43 Nuclear Assurance Corporation, *Near-Term Plutonium Market Outlook*, Norcross, Giorgia, USA, March 1983, p.12. (A Report, Ref: ORNL/Sub/83-40111/1, prepared for the US Department of Energy.) By 1982, 605 kgs. of plutonium recovered by BNFL from Italian magnox fuel – i.e. from Latina – had been allocated to the Superphénix

44 *New Scientist*, 25 July, 1985, p.17

45 *Hansard (Commons)*, 25 July 1962, col. 165w

46 *Hansard (Commons)*, 2 May 1985m col. 189w

47 *Nuclear Engineering International*, June 1982, p.13. *Hansard (Commons)*, 23 December, 1982, col. 621w

48 "Portugal confirms sale of uranium to Iraq", *Financial Times*, 28 March 1980

49 *Nucleonics Week*, 9 August 1990, p.3

50 *Nucleonics Week*, 4 October 1990, p.6

51 Nuclear Fuel, 20 August, 1990, p.8

52 Peter Lilley (Trade and Industry Secretary) quoted in *Nucleonics Week*, 8 August, 1991

Chapter 13

1   *Daily Mirror*, 29 May 1986

2   Gelder, R., J.S. Hughes, J.H. Mairs, K.B. Shaw, *Radiation Exposure Resulting from the Normal Transport of Radioactive Materials within the United Kingdom* (NRPB-R155), National Radiological Protection Board, Chilton, February 1984

3   Gelder, R., *The Radiological Impact of the Normal Transport of Radioactive Materials by Air* (NRPB-M219), National Radiological Protection Board, Chilton, 1990, p.11

4   Gelder, R., J.S. Hughes, J.H. Mairs, K.B. Shaw, op cit. p.34

5   Curren, W.D., and R.D. Bond (UKAEA, Winfrith), *Contamination Studies on pond-loaded flasks*, PATRAM 1980, IAEA, Vienna, 1980. p.850

6   Curtis, H.W., (NTL), *Experience of European LWR irradiated fuel transport*, PATRAM 1980, IAEA, Vienna, 1980, p.586

7   Curren, W.D., and R.D. Bond, op cit

8   Mairs, J.H., "The radiological impact of the normal rail transport of radioactive materials in the United Kingdom", *J. Soc. Radiol. Prot.* 3 (4) 1983

9   *Guardian*, 5 December, 1981

10 *SCRAM*, December 1989/January 1990, p.5

11 "Anger at nuclear waste near school", *Western Daily Press*, 3 February 1988

12 Hekstall-Smith, H.W., *Atomic Radiation Dangers – and what they mean to you*, J.M. Dent & Sons Ltd. London 1958, pp.84/5

13 Report of the committee to examine, *The Organisation for Control of Health and Safety in the United Kingdom Atomic Energy Authority*, Cmnd. 342, January 1958 para 15

14 *Living with radiation*, NRPB, Chilton, 1986. p.24

15 Gowing, M., *Independence and Deterrence: Britain and Atomic Energy 1945-1952*, Vol 1, Macmillan, London p.457 (footnote)

16 Flowers, B., *Nuclear Power and the Environment*, sixth report of the Royal Commission on Environmental Pollution, 1976, p.22

17 Clarke, R.H., "NRPB guidance on risk estimates and dose limits", *Radiological Protection Bulletin*, No. 88, January 1988
18 quoted in: Bunyard and Searle, "The effects of low-dose radiation", *Ecologist*, Vol 16 No 4/5 1986, p.171
19 *British Medical Journal*, 17 August 1985
20 Dr. Barry Lambert quoted in *Nucleonics Week*, 28 November 1985, p.9
21 *Nuclear Engineering International*, April 1990, p.2 (quoting Gardner et al, *British Medical Journal*, 17 February, 1990)

Chapter 14

1   Lohmann, D.H., *A review of the damage to packages from the Radiochemical Centre during transport*, PATRAM 1980, IAEA, Vienna. p.819
2   Hadjiantoniou, A., et al, *The Performance of Type A packaging under air crash and fire accident conditions*, PATRAM 1980, IAEA, Vienna, 1980, p.828
3   *Guardian*, 30 September 1988.
4   Fan et al *Survey of radioactive material transport in China*, PATRAM 1986, IAEA, Vienna. p.475
5   Bernado, B.C., *The need for emergency preparedness in transport of radioactive materials*, PATRAM 1980, IAEA, Vienna. p.1372
6   Grella, A.W., *A review of five years (1971–1975) accident experience in the USA involving nuclear transportation*, PATRAM 1976, IAEA, Vienna, p.235/6
7   *Nuclear Fuel*, 9 September 1985, p.3
8   Page, H., (BNFL) "United Kingdom experience of production of uranium fluorides", in: *Production of yellowcake and uranium fluorides* (proceedings of an advisory group meeting), IAEA, Vienna, 1980.
9   Chicken, J.C., E. Goldfinch, W.G. Milne, "Environmental impact of transporting radioactive materials" in: *The Environmental Impact of Nuclear Power*, published for the British Nuclear Energy Society by Thomas Telford Ltd. London 1981, p213
10 Shaw, K.B., J.S. Hughes, T.D. Gooding and L. McDonough, *Review of the Radiological Consequences Resulting from Accidents and Incidents Involving the Transport of Radioactive Materials in the UK from 1964 to 1988* (NRPB-M206), 1990, p.11
11 Goodridge, W., (UKAEA), "Review of Transport Accidents and Incidents involving radioactive materials", in: *The Safe Transport of Radioactive Materials*, Pergamon Press, Oxford, 1966, pp. 245 and 251
12 Shaw, K.B., J.S.Hughes, T.D.Gooding, L. McDonough, op cit. J.S. Hughes and K.B. Shaw *Radiological Consequences Resulting from Accidents and Incidents Involving the Transport of Radioactive Materials in the UK – 1989 Review*, NRPB, Chilton, 1990.
13 Grella, A.W., op cit. p.237/8
14 Grella, A.W., op cit. p.238/9
15 Bernado, B.C., op cit. p.1371
16 *Times*, 9 April 1968, p.1
17 Shaw, K.B., J.S.Hughes, T.D.Gooding, L. McDonough, op cit. p.20
18 Taylor, C.B.G., *Packaging and Transport of Radioisotopes*, paper published in

PATRAM 1976, IAEA, Vienna, p.76

19 Clarke, R., "Potential Consequences of Accidents" in: *The Urban Transportation of Irradiated Fuel*, Macmillan Press, London, 1984, pp.169-187
20 *Transport of Irradiated Fuel*, CEGB leaflet, August 1983
21 Perg report quoted in: North London Ecology Party, *Carrying the Can*, London, 1980
22 *The Windscale File*, Greenpeace, 1983, p.11
23 Wakstein, Charles, *Draft Study of the Consequences of a Magnox Flask Accident at Parliament Hill on the North London Line*, Report commissioned by Camden Alarm, May 1987
24 *Transport of Irradiated Fuel*, CEGB Leaflet, 1983
25 Cassidy, W., "Public attitudes to nuclear power: the role of local authority radiation monitoring", in *Uranium and Nuclear Energy*, Uranium Institute, London, 1989, p.298

Chapter 15

1 *Nucleonics Week*, 26 May 1988, p.11
2 Shaw, K.B., J.S. Hughes, T.D. Gooding and L. McDonough, *Review of the Radiological Consequences Resulting from Accidents and Incidents Involving the Transport of Radioactive Materials in the UK from 1964 to 1988* (NRPB-M206), NRPB, Chilton, 1990, paragraph 7.1.7, p.13
3 Taylor, C.B.G., *Packaging and Transport of Radioisotopes*, PATRAM 1976, IAEA, Vienna, 1976, p.77
4 "The boilermakers who caused a nuclear rethink", The Guardian, 1 July 1979
5 *Hansard (Commons)*, 27 July 1988, col. 247/8w
6 W. Patterson, *Nuclear Power*, Penguin Books, 1976. p.249
7 *Daily Express*, 17 May 1977
8 "Hunt goes on for danger van load", *Liverpool Echo*, 10 January 1991
9 "Is this why the Red spies were snooping", *Sunday Express*, 25 April 1976. (For a more detached view on this story – with more than a hint that it was hyped up to sell Pincher's book – see "Why the Great Spy Story fizzled out", *Times*, 26 April 1976 p.12.)
10 "Plutonium theft by terrorists a possibility, inquiry is told", *Times*, 22 June 1977, p.4
11 Kohn, Howard and Barbara Newman "How Israel got the nuclear bomb", *Rolling Stone*, 1 December, 1977, p.38
12 "U.S. Tightens Guard on Nuclear Facilities", *Washington Post*, 27 August 1985, p.5
13 Brown, M., P. Hague, D.J. Mather (UKAEA), "Criticality safety hazards arising from the transport of irradiated fuel elements" (final report of a contract between the Commission of the European Communities and the UKAEA), *European Appl. Res. Rept.-Nucl. Sci. Technol.* Vol 6, No. 3 (1985), p.459
14 "Citizen Smith strikes again!", *CUB* (Queen Mary College student union newspaper), December 1979, p.3
15 *Hansard (Commons)*, 27 July 1982, col. 438w
16 "Nuclear waste rail line a target", *Guardian*, 7 July 1982
17 "IRA vows to halt Belfast to Dublin railway link", *Times*, 8 May 1980

18   Philadelphia Inquirer, 27 April 1986, p.15
19   Goldschmidt, Bertrand, "Early Uranium Politics", in *Uranium and Nuclear Energy: 1984*, The Uranium Institute, London, 1985, p.322
20   *Washington Times*, 6 January 1986, p.5
21   *Washington Post*, 19 May 1986, p.16
22   Tombs, Francis, *A review of Nuclear Power in the United Kingdom*, pamphlet published by the Electricity Council, 1977
23   *Washington Times*, 26 August 1986, p.5

Chapter 16

1   *Hansard (Commons)*, 15 April 1981 col. 143w
2   quoted in "How to blow up a nuclear flask", *New Scientist*, 5 April 1984 p.5
3   *Hansard (Commons)*: 26 October 1989, col. 524w
4   *Hansard (Commons)*, 4 May 1989, cols. 187/8w
5   *Hansard (Commons)*, 23 May 1990 col. 186w
6   *Hansard (Commons)*, 8 February 1990, col. 782w
7   *Hansard (Commons)*, 3 July 1989, col. 26w
8   *Hansard (Commons)*, 24 July 1989, col. 459w
9   *Hansard (Commons)*, 1 November 1989, col. 224w
10   see, for example, NRC 1980 Annual Report, pp.117 and 119, published 1981
11   Jones, O.E., (Sandia Laboratories) "Advanced Physical Protection Systems for Nuclear Materials", published in *Safeguarding Nuclear Materials*, (proceedings of an IAEA symposium on the safegurading of nuclear materials), published by the IAEA, Vienna, 1976
12   For a cutaway diagram of an SST (provided by Sandia) see Ronald Allen Knieff, *Nuclear Energy Technology*, Hemishere Publishing Corporation (McGraw-Hill), Washington 1981, p.490
13   *Hansard (Commons)*, 3 February 1981, col. 75w
14   Yates, C., and E. Dewhurst, *Report on the Fire Test on a Plutonium Transport Container and recommendations*, 1960 (PRO ref: AB7 9802)
15   Fraser, D.C., and G. Jackson "An Investigation into the contamination of heavy vehicles carrying radioactive materials between Harwell, Windscale and Amersham", AERE HP/M 103, 1956 (quoted in Large & Associates, *Transportation of Irradiated Fuel*, Appendix 1)

Chapter 17

1   *Transport of CEGB Irradiated Nuclear Fuel*, CEGB, May 1983
2   NRPB, *Radiological Protection Bulletin*, No. 58, May 1984, p.10; NRPB, NAIR Handbook, 1987 edition. .p4
3   quoted in: *Local Authorities and Peacetime Emergency Planning*, a report by the Association of Metropolitan Authorities, November 1988, paragraph 3.1
4   Ibid
5   Charles Wakstein quoted in: *Carrying the Can*, The North London Ecology Party, 1980, p.22

6   *Hansard (Commons)*, 31 January 1983, col. 42/43w
7   Murdoch, W.S., "Site radiography incident involving a 20 curie cobalt-60 source", symposium paper published in *Handling of Radiation Accidents*, IAEA, Vienna, 1977
8   "Board Statement on Energency Reference Levels", in *Documents of the NRPB*, Vol 1, No. 4, 1990, p.18
9   Emergency Planning Guidance to Local Authorities, Home Office report, 1985, quoted in: *Civil Defence and Nuclear Accidents*, London Nuclear Information Unit, November 1986
10 "MOX fuel transport delayed", *Nuclear Engineering International*, September 1988, p.6
11 *Nuclear Engineering International*, April 1989, p.12; and September 1990, p.21
12 Curtis, H.W., and B. Lenail, *The transport of nuclear materials – the common link*, paper presented at the 12th annual symposium of the Uranium Institute, September 1987

Chapter 18

1   quoted in: *Sellafield News Letter*, #113, 25 July 1989, p.4
2   Mrs Thatcher, speaking on the 1pm news, Radio 4, 10.2.90
3   *Hansard (Commons)*, 3 April 1990, col 1024 (oral answers)
4   *ATOM*, May 1990 p.2
5   *Nuclear Fuel*, 21 May 1984, p.3; 13 August 1984, p.6
6   *BNFL Annual Report*, 1988/89, p.13
7   *UKAEA Press Release*, Dounreay, 16 February 1990
8   *ATOM*, May 1990, p.5
9   *ATOM*, August 1988, p.13
10 *Nuclear Fuel*, 11 October 1990, p.11
11 *Nuclear Fuel*, 28 May 1990, p.2
12 *Nuclear Engineering International*, February 1989, p.11
13 *ATOM*, August 1988 p.15
14 Helliwell, P., and J. Spink, *The Transport of Plutonium by Air and Sea*, European Proliferation Information Centre, London, 1989, p.89
15 First Report from the Select Committee on the Environment, 1985/86, p.cxxiii
16 "BR safety move on nuclear cargoes", *Guardian*, 21 May 1987
17 *Hansard (Commons)* 21 October 1985, col. 90w
18 Curie, Eve, *Madame Curie*, William Heinemann Ltd. London. 1938, p.233
19 Chicken, J.C., *Summary of the risk assessment made of the transport of plutonium nitrate*, PATRAM 1980, IAEA, Vienna, p.914

Index